Deterrence and First-Strike Stability in Space

A Preliminary Assessment

Forrest E. Morgan

Prepared for the United States Air Force

PROJECT AIR FORCE

The research described in this report was sponsored by the United States Air Force under Contract FA7014-06-C-0001. Further information may be obtained from the Strategic Planning Division, Directorate of Plans, Hq USAF.

Library of Congress Cataloging-in-Publication Data

Morgan, Forrest E.
  Deterrence and first-strike stability in space : a preliminary assessment / Forrest E.
  Morgan
    p. cm.
  Includes bibliographical references.
  ISBN 978-0-8330-4913-1 (pbk. : alk. paper)
  1. Space warfare--Government policy--United States. 2. Space weapons--United States.
  3. Deterrence (Strategy). I. Title.

  UG1523.M587 2010
  358'.8--dc22

                                                                    2010008354

The RAND Corporation is a nonprofit research organization providing objective analysis and effective solutions that address the challenges facing the public and private sectors around the world. RAND's publications do not necessarily reflect the opinions of its research clients and sponsors. **RAND®** is a registered trademark.

Published 2010 by the RAND Corporation
1776 Main Street, P.O. Box 2138, Santa Monica, CA 90407-2138
1200 South Hayes Street, Arlington, VA 22202-5050
4570 Fifth Avenue, Suite 600, Pittsburgh, PA 15213-2665
RAND URL: http://www.rand.org/
To order RAND documents or to obtain additional information, contact
Distribution Services: Telephone: (310) 451-7002;
Fax: (310) 451-6915; Email: order@rand.org

Preface

The United States now, more than at any point in its history, depends on space systems for its national security—and much more so than any other country. This, combined with the fact that those systems are becoming vulnerable to a growing number of potential adversaries, suggests that first-strike stability in space is eroding. Consequently, leaders in the U.S. defense community and particularly those in Air Force Space Command have asked the following questions: Can future enemies be deterred from attacking U.S. space systems? To what degree is deterrence reliable, and under what circumstances might it fail? What can the United States do to fashion the most robust space deterrence regime and strengthen first-strike stability in space?

This monograph provides a preliminary examination of these questions and develops a framework for further analysis. It begins with a historical review of the shifting dynamics of first-strike stability in space and explains why that stability may now be in peril. Then, it applies the principles of deterrence to the strategic environment of space to identify the unique challenges presented there and to illustrate why it may be difficult to deter future adversaries from attempting to degrade or destroy U.S. space capabilities in certain scenarios. Finally, it proposes a framework for a comprehensive national space deterrence strategy and identifies the areas in which future research will be needed to determine the optimal mix of policies, strategies, and systems for establishing the most effective and affordable deterrence regime.

This monograph will be of interest to officials in the U.S. defense community who are tackling these important questions, as well as to scholars and analysts engaged in the study of the changing threat environment and challenges to U.S. space deterrence.

The research reported here was prepared for a fiscal year 2009 study, "Space Deterrence." The work was conducted within RAND Project AIR FORCE (PAF) on a PAF-wide basis with oversight provided by PAF's Strategy and Doctrine Program. Related research that may interest readers of this monograph includes *Dangerous Thresholds: Managing Escalation in the 21st Century*, by Forrest E. Morgan, Karl P. Mueller, Evan S. Medeiros, Kevin L. Pollpeter, and Roger Cliff (MG-614-AF).

RAND Project AIR FORCE

RAND Project AIR FORCE (PAF), a division of the RAND Corporation, is the U.S. Air Force's federally funded research and development center for studies and analyses. PAF provides the Air Force with independent analyses of policy alternatives affecting the development, employment, combat readiness, and support of current and future aerospace forces. Research is conducted in four programs: Force Modernization and Employment; Manpower, Personnel, and Training; Resource Management; and Strategy and Doctrine.

Additional information about PAF is available on our Web site: http://www.rand.org/paf/

Contents

Figures

Summary

Space stability is a fundamental U.S. national security interest. Unfortunately, that stability may be eroding. Since the end of the Cold War, U.S. military forces have repeatedly demonstrated their dominance in conventional warfare, and future enemies will be well aware that the dramatic warfighting advantage that U.S. forces possess is largely the result of support from space. With a growing number of states acquiring the ability to degrade or destroy U.S. space capabilities, the probability that space systems will come under attack in a future crisis or conflict is ever increasing. Deterring adversaries from attacking some U.S. space systems may be difficult due to these systems' inherent vulnerability and the disproportionate degree to which the United States depends on the services they provide. Nevertheless, the United States can fashion a regime to raise the thresholds of deterrence failure in terms of destructive attacks on its space systems and thus achieve a measure of first-strike stability in space during crises and at some levels of limited war. (See pp. 7–16.)

Estimated Thresholds of Space Deterrence Failure

While the factors above suggest that stability in space is eroding, it would be overly simplistic to assume that the thresholds of deterrence failure are the same for all space systems or at all levels of confrontation. In any given crisis or conflict, an adversary would have to weigh a range of factors in contemplating attacks on U.S. space capabilities. The risks incurred or benefits expected in a space attack would vary greatly in the context of any specific scenario. Consequently, it is less a question of whether would-be aggressors can be deterred from attacking U.S. space systems than of what kinds of attacks against which capabilities could be deterred under what circumstances. (See pp. 16–21.)

As Figure S.1 illustrates, an adversary's assessment of the costs and benefits of attacking a U.S. space system would likely vary from one prospective target set to another at each level of conflict, and the threshold of deterrence failure would be different for nondestructive attacks (i.e., "reversible-effects" attacks) than for destructive attacks (those that cause damage). (See pp. 16–21.)

Some of these thresholds are quite low today. An opponent in a confrontation with the United States that has not yet engaged in conventional terrestrial hostilities might consider reversible-effects attacks on U.S. space-based intelligence, surveillance, and reconnaissance (ISR) and communication assets to be a promising means of degrading the United States' ability to respond to the crisis, with relatively low risk of serious retribution compared to that of a destructive attack on one or more U.S. satellites. Fearing the onset of U.S. air strikes, the adversary might also begin jamming Global Positioning System (GPS) signals in areas around command-and-control nodes and other important facilities to degrade the accuracy of U.S. precision-guided weapons. Even after fighting has begun, a savvy adversary might continue to abstain from destroying U.S. satellites in a limited war for fear of escalating the conflict, particularly if the reversible-effects attacks continued to yield comparable levels of benefit. However, should the terrestrial conflict escalate, it would become increasingly difficult to deter an enemy with the appropriate capabilities from carrying out destructive attacks in space. At some point, the conflict would likely reach a threshold at which the growing benefits of transitioning to destructive attacks on certain space systems would overtake the dwindling costs of doing so. In fact, satellites used for reconnaissance and ocean surveillance—being high-value, low-density assets—might become targets even at relatively low levels of conflict, and the adversary might attempt to damage dedicated U.S. military satellite communication (MILSATCOM) assets as well. (See pp. 16–21.)

Conversely, since commercial satellite communication (SATCOM) platforms typically support a host of international users as well as U.S. forces, the political costs and escalatory risks of carrying out destructive attacks on those assets might deter the opponent from attempting to do so until the conflict escalated to a higher level. Satellites supplying positioning, navigation, and timing (PNT) data—i.e., GPS— would probably be relatively safe from destructive attack until very high levels of conflict, because the distributed nature of that system would make it difficult for an opponent to realize much benefit from individual attacks. The adversary would also likely be deterred from damaging U.S. satellite early-warning system (SEWS) assets to avoid risking inadvertent escalation to the nuclear threshold, but that firebreak would almost certainly collapse with the conclusion that such escalation is inevitable and that it is in the adversary's interest to launch a preemptive nuclear strike. Alternatively, because the strategic surveillance and warning system also supports efforts to locate and destroy mobile

Figure S.1
Notional Space Deterrence Capabilities, by System Type at Various Levels of Conflict

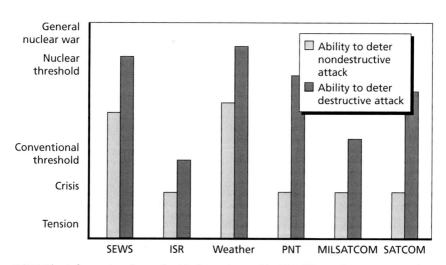

NOTE: The information shown in this figure is provided for illlustrative purposes only and is not based on an analysis of empirical data.
RAND *MG916-S.1*

conventional missile launchers, the adversary might risk dazzling SEWS satellites at lower levels of conflict to impede U.S. efforts to locate and destroy those launchers. (See pp. 16–21.)

Weather satellites, surprisingly, might be the space assets that are safest from attack. Attacking assets supporting the highly globalized international meteorological system would result in considerable political costs, and the robust infrastructure supporting that system would limit the benefits of individual attacks against it. (See pp. 16–21.)

Space Deterrence and General Deterrence

Although this assessment focuses specifically on space deterrence and first-strike stability in space, it is important to appreciate the interdependencies between these factors and general deterrence and stability writ large. Given the extent to which space support enhances U.S. conventional military capabilities, an adversary weighing the risks and potential benefits of war with the United States might be encouraged toward greater aggression by the belief that attacking space systems would degrade U.S. warfighting capabilities enough to enable the attainment of objectives at acceptable costs. As a result, weaknesses in space deterrence can undermine general deterrence. Conversely, if a prospective adversary concludes that the probable cost-benefit outcome of attacking U.S. space systems is unacceptable, it is forced to weigh the risks and benefits of aggressive designs in the terrestrial domain against the prospect of facing fully capable, space-enhanced U.S. military forces. In sum, effective space deterrence fortifies general deterrence and stability. (See p. 21.)

Deterrence in the Space Environment

Deterrence entails discouraging an opponent from committing an act of aggression by manipulating the expectation of resultant costs and benefits. Deterring attacks on U.S. space systems will require the United States to fashion credible threats of punishment against potential opponents, persuade adversaries that they can be denied the benefits of their aggression, or some combination of both approaches. However, fashioning a space deterrence regime that is sufficiently potent and credible will be difficult given that U.S. warfighting capabilities, much more so than those of any potential adversary, depend on space support. Threatening to punish aggressors by destroying their satellites might not deter them from attacking U.S. assets—a game of satellite tit-for-tat would likely work to the adversary's advantage. Conversely, threats of punishment in the terrestrial domain may lack credibility in crises and at lower levels of limited war and would likely be irrelevant at higher levels of war, when heavy terrestrial attacks are already under way. Denial strategies face other hurdles. Efforts to deny adversaries the benefits of space aggression are hindered by the inherent vulnerability of some important U.S. space systems and the high degree of U.S. dependence on those assets. As long as those systems are vulnerable, the enemy's benefit in attacking space assets is proportionate to the United States' dependence on the capabilities they provide. (See pp. 24–33.)

The Task Is Not Impossible

While these factors suggest that it may be difficult to deter potential enemies from attacking certain U.S. space systems in some circumstances, the task of strengthening first-strike stability in space is by no means impossible. As illustrated earlier, the orbital infrastructures of some U.S. systems are already sufficiently robust that they present poor targets for prospective attackers. The challenge will be to find ways to raise the thresholds of deterrence failure for those systems that are both vulnerable and important for force enhancement. Meeting this challenge will require the United States to develop and employ a coherent national space deterrence strategy. (See p. 35.)

The Need for a National Space Deterrence Strategy

The United States can raise the thresholds of deterrence failure in crises and at some levels of limited war by implementing a coordinated national space deterrence strategy designed to operate on both sides of a potential adversary's cost-benefit decision calculus simultaneously. The foundation and central pillar of such a strategy would be a national space policy that explicitly condemns the use of force in space and declares that the United States will severely punish any attacks on its space systems and those of friendly states in ways, times, and places of its choosing. Cognizant of the fundamental U.S. interest in space stability, such a policy would embrace diplomatic engagement, treaty negotiations, and other confidence-building measures, both for whatever stabilizing effects can be attained from such activities and because demonstrating leadership in these venues helps to characterize the United States as a responsible world actor with the moral authority to use its power to protect the interests of all spacefaring nations. In these settings and others, all U.S. policies, statements, and actions should be carefully orchestrated to bolster already emerging international taboos on space warfare and enhance the credibility of U.S. threats to punish space aggressors in multiple dimensions—in the terrestrial and informational domains as well as in space, through diplomatic and economic means, in addition to the use of force. Such an approach would raise the potential costs in ways that future opponents would have to factor into their decision calculations in any crisis in which they are tempted to attack orbital assets. (See pp. 37–44.)

At the same time, the United States should engage in a comprehensive and coordinated effort to persuade potential adversaries that the probability of obtaining sufficient benefit from attacking space assets would not be high enough to make it worth suffering the inevitable costs of U.S. retribution. Part of such a strategy would entail perception management: The United States should, to the greatest extent possible, conceal vulnerabilities of its space systems and demonstrate the ability to operate effectively without space support. However, perception management can only go so far in the face of observable weaknesses. Therefore, the strategy should also pursue multiple avenues to make vulnerable U.S. space systems more resilient and defendable, thereby demonstrating tangible capabilities to deny potential adversaries the benefits of attacking in space. (See pp. 44–45.)

Although satellites are inherently difficult to defend, there are a variety of options that the United States should explore for reducing the vulnerabilities of its space systems. Possibilities include making greater investments in passive defenses, exploring approaches to active defenses, dispersing capabilities across a larger number of orbital platforms, and developing terrestrial backups to space support. It may also be beneficial to disperse some U.S. national security payloads onto satellites owned by a range of other nations and business consortia friendly to the United States and also to engage in data-sharing arrangements with them. Such approaches would create an international security space infrastructure that is more robust than the sum of its individual systems, raise escalation risks for anyone contemplating attacks on that infrastructure, and strengthen international support for U.S. threats of punishment in response to attacks. (See pp. 45–48.)

Current deficiencies in space situational awareness (SSA) are sources of particular concern. While many options exist for punishing space aggressors and reducing the benefits of their attacks, nearly all of them depend to some degree on improvements in SSA. Poor SSA undermines the credibility of threats of punishment in some scenarios, as the attacker may expect to have a reasonable chance of striking anonymously. All active defenses require better SSA than what current capabilities provide, and many passive defenses could also be improved with better SSA. Improving SSA should be one of the United States' top priorities in its efforts to develop the capabilities needed for an effective space deterrence regime. (See p. 48.)

A Way Forward

Although this monograph proposes the broad outlines of a comprehensive space deterrence regime, more work is needed to evaluate which of the various options discussed here are viable and what combination would best support a reliable strategy. Such work would consist of an integrated analysis of the space deterrence problem as a complex system and an examination of the consequences of alternative courses of action, both by the United States and by its most likely potential adversaries, across a range of scenarios. Insights gained from such an examination would inform further analyses to determine near- and far-term approaches for achieving the optimal mix of policies, strategies, and systems for establishing the most effective and affordable deterrence regime. (See pp. 51–53.)

This effort would entail a broad investigation, bringing a wide range of analytical methods to bear and ultimately integrating technical assessments with expert judgment. Planners would need to gather a good deal of information, but much of it is available from intelligence sources or has already been developed in previous studies. The investigation would begin with surveys of that work. Risk analyses and engineering assessments would play important roles in determining degrees of vulnerability and the most promising approaches for mitigating them. Crisis-gaming and war-gaming would be essential tools for exploring the dynamics of deterrence and stability across a range of scenarios and levels of conflict. Other aspects of the investigation would include an examination of space law and consultation with space experts in the U.S. analytical community and elsewhere. Ultimately, having gathered the findings of the surveys, assessments, and analyses, planners would be able to refine and further develop the comprehensive space deterrence strategy outlined in this monograph and offer recommendations for its implementation. (See pp. 51–53.)

Acknowledgments

I would like to express my sincerest appreciation to reviewers Bruce W. MacDonald, Lara S. Schmidt, and James Thomson for their help in developing this monograph. Their prompt and thorough reviews, thoughtful analyses, and frank recommendations improved the end result substantially. I would also like to thank RAND colleagues Paul Deluca, Andrew Hoehn, Martin Libicki, Karl Mueller, Richard Mesic, David Ochmanek, Chad Ohlandt, Jan Osburg, and Carl Rhodes for their insightful comments and recommendations. Their assistance also improved the quality of this monograph. Finally, special thanks go to Maria Falvo and Laura Novacic for their prompt and efficient administrative assistance in preparing the draft and to Lauren Skrabala for her diligent work in editing it.

Abbreviations

ASAT	antisatellite
BMD	ballistic missile defense
EMP	electromagnetic pulse
FOBS	fractional orbital bombardment system
GPS	Global Positioning System
ISR	intelligence, surveillance, and reconnaissance
LEO	low earth orbit
MAD	mutual assured destruction
MILSATCOM	military satellite communication
OODA	observe, orient, decide, act
PNT	positioning, navigation, and timing
RF	radio frequency
SALT	Strategic Arms Limitation Treaty
SATCOM	satellite communication
SEWS	satellite early-warning system
SSA	space situational awareness

CHAPTER ONE

Introduction

Given the great extent to which the United States depends on space systems for its national security and economic prosperity, U.S. policymakers and military leaders are becoming increasingly concerned that future adversaries might attack those systems. U.S. military forces operate in distant theaters and employ ever more sophisticated equipment and doctrines that rely on advanced surveillance, reconnaissance, communication, navigation, and timing data, most of which is produced or relayed by satellites. The ground infrastructure that supports these assets has long been vulnerable to attack, and a growing number of states now possess or are developing means of attacking satellites and the communication links that connect them to users and control stations. Due to the dramatic warfighting advantage that space support provides to U.S. forces, security analysts are nearly unanimous in their judgment that future enemies will likely attempt to "level the playing field" by attacking U.S. space systems in efforts to degrade or eliminate that support. All of this suggests that first-strike stability in space may be eroding.

First-strike stability is a concept that Glenn Kent and David Thaler developed in 1989 to examine the structural dynamics of mutual deterrence between two or more nuclear states.[1] It is similar to crisis stability, which Charles Glaser described as "a measure of the countries' incentives not to preempt in a crisis, that is, not to attack first

---

[1] Glenn A. Kent and David E. Thaler, *First-Strike Stability: A Method for Evaluating Strategic Forces*, Santa Monica, Calif.: RAND Corporation, R-3765-AF, 1989.

in order to beat the attack of the enemy,"[2] except that it does not delve into the psychological factors present in specific crises. Rather, first-strike stability focuses on each side's force posture and the balance of capabilities and vulnerabilities that could make a crisis unstable should a confrontation occur.[3]

Space stability issues differ from the Kent-Thaler conception of first-strike stability in that nuclear forces are not directly involved, so the risk of prompt catastrophic damage in the event of a deterrence failure is not nearly as great. However, several other strong parallels exist between first-strike stability in space and in the nuclear realm. First, space support substantially enhances operational warfighting capabilities in the terrestrial domain that are threatening to potential enemies. At the same time, satellites are difficult to defend against adversaries with capabilities to attack them. As a result, space, like the nuclear realm, is an offense-dominant environment with substantial incentives for striking first should war appear probable. Second, deterrence failures in space, though not as immediately catastrophic as nuclear deterrence failures, could, nonetheless, be very costly given the resources invested in orbital infrastructure and the many security and economic functions that benefit from space support. And, like nuclear deterrence failures, the costs of warfare in space would likely be shared by third parties due to global economic interdependence and multinational ownership of many space systems—all the more so if kinetic attacks on satellites litter important orbits with debris. Finally, there is a parallel between nuclear and space deterrence in that significant thresholds are perceived in both realms, the crossing of which could lead to reprisals, follow-on attacks, and rapid escalation.[4]

---

[2] Charles L. Glaser, *Analyzing Strategic Nuclear Policy*, Princeton, N.J.: Princeton University Press, 1990, p. 45.

[3] Kent and Thaler, 1989, p. 2.

[4] For more on thresholds and escalation risks in the current security environment, see Forrest E. Morgan, Karl P. Mueller, Evan S. Medeiros, Kevin L. Pollpeter, and Roger Cliff, *Dangerous Thresholds: Managing Escalation in the 21st Century*, Santa Monica, Calif.: RAND Corporation, MG-614-AF, 2008.

While strategic thinkers largely agree that U.S. space systems present tempting targets for future adversaries, there is wide debate on what to do about this threat. Some argue that, due to the difficulty of defending orbital assets from capable attackers, the United States should simply attempt to discourage hostile actors from attacking satellites by continuing to promote the international norm that space should be preserved as a sanctuary from war.[5] At the other end of the spectrum are those who argue that the United States should arm itself for offensive and defensive counterspace operations and, in the event of war, protect its space assets by forcefully dominating the medium in a fashion similar to the way it has in other domains.[6] Positions that fall between these extremes include arguments for more passive and active defenses, dispersal of space capabilities, and developing alternative means of support, thereby reducing U.S. dependence on space.

Unfortunately, many of these arguments miss an important point: Given that the United States benefits so much from uninterrupted access to space support, a fundamental U.S. national security interest in space—perhaps the most important one—is stability. Granted, developing the ability to defend U.S. space assets is an important objective, and should the United States find itself at war with an adversary whose warfighting capabilities are substantially enhanced by space systems, U.S. military leaders would likely want the ability to deny those adversaries access to space support. However, once the threshold of destructive attacks against satellites is crossed, the United States and its allies may suffer high costs even if they ultimately "win" the space engagement and dominate that domain. Such costs would not be limited to

---

[5] See, for instance, Bruce M. Deblois, "Space Sanctuary: A Viable National Strategy," *Airpower Journal*, Vol. 12, No. 4, Winter 1998, and David W. Ziegler, "Safe Heavens: Military Strategy and Space Sanctuary," in Bruce M. DeBlois, ed., *Beyond the Paths of Heaven: The Emergence of Space Power Thought by the School of Advanced Airpower Studies*, Maxwell AFB, Ala.: Air University Press, September 1999.

[6] See Everett C. Dolman, *Astropolitik: Classical Geopolitics in the Space Age*, London: Frank Cass, 2002; Colin S. Gray and John B. Sheldon, "Spacepower and the Revolution in Military Affairs: A Glass Half Full," in Peter L. Hays, James M. Smith, Alan R. Van Tassel, and Guy M. Walsh, eds., *Spacepower for a New Millennium*, New York: McGraw-Hill, 2000; and Simon P. Worden, "Space Control for the 21st Century: A Space 'Navy' Protecting the Commercial Basis of America's Wealth," in Hays et al., 2000.

the orbital infrastructure, because economic functions and terrestrial military operations would also likely suffer from degradations in space support. In fact, because the United States enjoys greater economic, scientific, and national security benefits from its space systems than any other state, it has the most to lose in conflicts in the space domain. Grasping the significance of this situation, the recently released final report from the bipartisan congressional commission appointed to review the strategic posture of the United States recommended that the nation "[d]evelop and pursue options for advancing U.S. interests in stability in outer space and in increasing warning and decision-time."[7] As Bruce MacDonald recently testified before the Strategic Forces Subcommittee of the House Armed Forces Committee,

> Our overall goal should be to shape the space domain to the advantage of the United States, and to do so in ways that are stabilizing and enhance U.S. security. The U.S. has an overriding interest in maintaining the safety, survival, and function of its space assets so that the profound military, civilian, and commercial benefits they enable can continue to be available to the United States and its allies.[8]

With these concerns in mind, this monograph examines an issue that has recently become prominent in the space strategy debate—that is, whether the United States can establish an effective regime to deter potential enemies from attacking its space systems.[9] It further considers

---

[7]   William J. Perry, James R. Schlesinger, Harry Cartland, John Foster, John Glenn, Mortin Halperin, Lee Hamilton, Fred Ilke, Keith Payne, Bruce Tarter, Ellen Williams, and James Woolsey, *America's Strategic Posture: The Final Report of the Congressional Commission on the Strategic Posture of the United States*, Washington, D.C.: United States Institute of Peace Press, 2009, p. 71.

[8]   Bruce W. MacDonald, testimony before the Strategic Forces Subcommittee, House Armed Forces Committee, March 18, 2009.

[9]   For other recent important contributions to the national debate, see Thomas G. Behling, "Ensuring a Stable Space Domain for the 21st Century," *Joint Force Quarterly*, No. 47, 4th Quarter 2007; Bruce W. MacDonald, *China, Space Weapons, and U.S. Security*, Washington, D.C.: Council on Foreign Relations, September 2008; Robert Butterworth, "Fight for Space Assets, Don't Just Deter," Policy Outlook, Washington, D.C.: George C. Marshall Insti-

what concert of actions might contribute to the enhancement of first-strike stability in space and, conversely, whether certain actions might further erode it.

Strengthening first-strike stability in space could be a tough challenge given the nature of the domain and the extent to which the United States depends on vulnerable systems there. To put the problem in perspective, we must first consider how, over the history of U.S. space operations, the emphasis has shifted from supporting national strategic missions almost exclusively in the early years to enabling U.S. conventional military dominance in the post–Cold War era. At the same time, there has been a shift from a period when satellites, though inherently fragile, were relatively isolated from threats due to the inability of most adversaries to reach them, to the present condition in which continued satellite fragility, coupled with the spread of space weapon technology, is creating a distinct first-strike advantage that could manifest as a surprise attack in space against selected U.S. systems at the onset of a future conflict.

With that foundation laid, this monograph examines the fundamentals of deterrence theory and determines how those principles play out in the space strategic environment. Deterring attacks on space systems will require the United States to fashion credible threats of punishment against potential opponents, develop a demonstrable ability to deny them the benefits of attack, or some combination of both approaches. But, as this monograph explains, fashioning deterrent threats that are sufficiently potent and credible will be difficult given the fact that U.S. warfighting capabilities, much more so than those of any potential adversary, depend on space support.

Nevertheless, this monograph argues that the United States can raise the thresholds of deterrence failure in terms of destructive attacks on its space systems and thus restore a measure of first-strike stability

tute, November 2008; Michael Kreppon, testimony before the Strategic Forces Subcommittee, House Armed Forces Committee, March 10, 2009; and Roger G. Harrison, Darin R. Jackson, and Collin G. Shackelford, *Space Deterrence: The Delicate Balance of Risk*, Colorado Springs, Colo.: Eisenhower Center for Space and Defense Studies, April 2009. Recent RAND work on space deterrence issues includes research by Russell D. Shaver and Richard Mesic.

in space in crises and at some levels of limited war. However, such a regime cannot be based solely on what most people envision when they think of deterrence: threats of retribution alone. Rather, effective deterrence in space will require a coordinated national strategy designed to operate on both sides of a potential adversary's cost-benefit decision calculus simultaneously. Such a strategy would raise the potential costs of attacking U.S. space systems by threatening a range of punitive responses in multiple domains while at the same time reducing the benefits of enemy attacks by improving defenses, dispersing and concealing space capabilities, reducing U.S. dependence on space by developing alternative capabilities, and demonstrating the ability to rapidly replenish whatever losses are sustained.

This monograph provides an initial template for such a strategy and a menu of options for strengthening deterrence by making U.S. space capabilities more robust. However, readers are cautioned that a logical template is not a strategy, and a list of options says nothing about what combination of options is most viable and affordable. Additional study will be needed to better understand the complex dynamics that might determine success or failure of deterrence in space and to develop more detailed recommendations for the U.S. Air Force and the nation. Therefore, the monograph concludes with a proposal that Air Force Space Command sponsor a thorough systems analysis of the space deterrence problem to further identify the optimal mix of policies, strategies, and systems for establishing the most effective and affordable deterrence regime.

CHAPTER TWO

The Shifting Dynamics of Stability in Space

The Historical Backdrop

The dawn of the space age occurred nearly simultaneously with the dawn of the nuclear age, and as both of them emerged in a geopolitical context of Soviet-American rivalry, U.S. policies on what space capabilities would be developed, how they would be employed, and how they would be portrayed to domestic and international audiences were profoundly shaped by Cold War exigencies. Because nearly all early national security space capabilities were developed to support nuclear deterrence and nuclear warfighting missions, first-strike stability in space was inexorably tied to crisis stability between the superpowers. The early years of the space age were dangerous times, as Moscow attempted to deploy offensive missiles to Cuba and threatened to place nuclear weapons in orbit.

During that era, however, neither the Soviet Union nor the United States had any ability to attack satellites except by the crudest means—namely, launching nuclear weapons into space. Ultimately, the Cuban Missile Crisis frightened Cold War leaders, making them more cautious. The superpowers settled into a stable relationship as Washington and Moscow realized that they were locked in a condition of mutual assured destruction (MAD). This stability initially extended into space, but it was not to last in that domain. As the Cold War entered its final stages, the superpowers increasingly found ways to use space capabilities to support their conventional military forces, and Moscow began testing a co-orbital antisatellite (ASAT) and ground-based directed-energy ASAT weapons. As a result, U.S. leaders began worrying about how to protect their orbital infrastructure, and they ordered work to begin on their own counterspace systems. Those programs lost momentum with the demise of bipolar tensions at the end of the Cold War; however, concerns about space system vulnerabilities returned in the post–Cold War environment as the United States repeatedly demonstrated its space-enabled conventional military dominance and other states began developing ways to degrade or destroy U.S. space capabilities.

The Period of Strategic Uncertainty

In the mid-1950s, with the United States racing the Soviet Union in the development of rocket and satellite programs, the U.S. Air Force had visions of fielding orbital space planes that would be capable of performing space analogs to the air superiority and strategic bombardment missions that it had traditionally conducted so effectively in the terrestrial environment.[1] However, due to the critical need to collect strategic intelligence in the vast regions of the Soviet interior denied to aerial reconnaissance, the Eisenhower administration's first priority was to get the international community to accept the legality of "freedom of space" for reconnaissance satellite overflight.[2] Moscow set a precedent for this principle when Soviet leaders declared their own right to freedom of space with the launch of Sputnik in October 1957. But soon afterward, the United Nations Committee for the Peaceful Uses of Outer Space issued an opinion that such freedom applied only to spacecraft on peaceful missions. Consequently, in 1958, as part of a coordinated campaign to characterize all U.S. satellite programs as

---

[1]   Adam L. Gruen, "Manned Versus Unmanned Space Systems," in R. Cargill Hall and Jacob Neufeld, eds., *The U.S. Air Force in Space: 1945 to the 21st Century*, Proceedings of the Air Force Historical Foundation Symposium, Andrews AFB, Md., September 21–22, 1995, Washington, D.C.: Air Force History and Museums Program, 1998, pp. 70–71; Curtis Peebles, *High Frontier: The United States Air Force and the Military Space Program*, Washington, D.C.: Air Force History and Museums Program, 1997, pp. 16–22.

[2]   Walter A. McDougall, *The Heavens and the Earth: A Political History of the Space Age*, Baltimore, Md.: Johns Hopkins University Press, 1997, pp. 115–118; General Bernard A. Schriever, "Military Space Activities: Recollections and Observations," in Hall and Neufeld, 1998, p. 14.

peaceful uses of space, the U.S. Department of Defense prohibited the military services from developing (or even publicly mentioning) any kind of orbital weapons.[3]

Most U.S. and Soviet national security space systems developed in the 1960s were dedicated to supporting nuclear warfighting and deterrence missions. Space-based strategic reconnaissance was followed by weather satellites to support mission planning and communication satellites to provide survivable command and control of nuclear forces. By the early 1970s, the United States was operating a satellite early-warning system (SEWS) to detect missile launches and nuclear detonations, and the Soviet Union was attempting to develop a system with comparable capabilities. As this orbital infrastructure developed, the specter of nuclear war in and from space emerged when Nikita Khrushchev, on several instances before and during the Cuban Missile Crisis, threatened to field a fractional orbital bombardment system (FOBS), ultimately prompting the Kennedy and Johnson administrations to approve the development of nuclear-armed ASAT interceptors.[4] As a result, the U.S. Air Force's Program 437 stood alert on Johnson Island in the Pacific Ocean for several years in the late 1960s, as did the U.S. Army's Program 505 on Kwajalein Atoll, but by then the threat was already diminishing.[5] In October 1963, Moscow agreed to United Nations General Assembly Resolution 1884 (XVIII) banning the placement of all weapons of mass destruction in orbit around the earth. That prohibition was further cemented in 1967 when the United States, United Kingdom, and Soviet Union signed and ratified the

---

[3] McDougall, 1997, p. 185; Peebles, 1997, pp. 10–11.

[4] Paul B. Stares, *The Militarization of Space: U.S. Policy, 1945–1984*, Ithaca, N.Y.: Cornell University Press, 1985, pp. 99–100. In the FOBS concept, the Soviets envisaged launching nuclear-armed satellites into orbit in a southerly direction, then de-orbiting them short of one full revolution—with the weapons striking targets in the United States. The strategic rationale for such a system was that the weapons would approach the North American continent from the south, thereby evading U.S. missile warning radars, which, in that era, all faced north.

[5] Clayton K. S. Chun, *Shooting Down a "Star": Program 437, the US Nuclear ASAT System and Present-Day Copycat Killers*, Center for Aerospace Doctrine Research and Education Paper No. 6, Maxwell AFB, Ala.: Air University Press, April 2000.

Treaty on Principles Governing Activities of States in the Exploration and Uses of Outer Space, Including the Moon and Other Celestial Bodies, commonly known as the Outer Space Treaty.[6] Despite signing the agreement, the Soviets began testing a FOBS that same year, maintaining the letter, if not the spirit, of the treaty's constraints by not putting nuclear devices on any of its test vehicles. They began testing a co-orbital ASAT system the following year.[7] By this time, however, an ever-increasing number of policymakers and security analysts on both sides of the Iron Curtain had realized that strategic arsenals and survivable second-strike capabilities had grown to the point that nuclear war between the superpowers could not be fought without unacceptable costs, and a sense of imposed stability began to settle over the strategic environment. This stability, along with several other factors, contributed to the temporary thaw in U.S.-Soviet relations in the early to mid-1970s, characterized as détente.[8]

The Illusion of Sanctuary

With the space and nuclear deterrence missions so closely integrated in both the United States and the Soviet Union, the stability that MAD imposed on the strategic environment extended into the space domain. In late 1971, Moscow suspended FOBS and ASAT weapons testing. Within a year, U.S. and Soviet negotiators reached agreement on the Strategic Arms Limitation Treaty (SALT) and the Anti-Ballistic Missile Treaty. These accords represented efforts to reduce pressures for further investment in strategic weapons and establish in international law formal mechanisms for crisis management. SALT was the first of

---

[6]   The Treaty on Principles Governing the Activities of States in the Exploration and Use of Outer Space, Including the Moon and Other Celestial Bodies was signed in London, Moscow, and Washington on January 27, 1967, and ratified in the U.S. Senate on April 25, 1967. It entered force on October 10, 1967.

[7]   Stares, 1985, p. 99.

[8]   The other factors included growing antipathy between the Soviet Union and China, which led the leaders of those countries to curry Washington's favor as a foil against each other, and the Nixon administration's successful exploitation of those developments.

several agreements to designate satellites as "national technical means" for treaty verification.[9] The fact that each signatory pledged not to interfere with the national technical means of the other indicated that Washington and Moscow had recognized that attacking each other's space assets could be destabilizing. Throughout the remainder of the Cold War, although isolated incidents of harassment occurred in the form of reversible-effects attacks such as "dazzling," neither antagonist risked destructive attacks on satellites supporting the other's nuclear forces for fear that such acts might be interpreted as the first step in a surprise nuclear war. Over time, such stability concerns and treaty prohibitions convinced some U.S. analysts that space could be preserved as a weapon-free sanctuary.[10]

But détente was not to last. In 1975, the Soviets began testing ground-based lasers and other directed-energy weapons, dazzling three U.S. satellites in multiple incidents; the following year, they resumed experimenting with co-orbital ASAT systems. These developments, coupled with the Soviets' increasing ability to use space-based assets to support their conventional military forces, were sufficiently troubling to the Ford administration that the President was persuaded, just before leaving office in January 1977, to order that work begin on a new U.S. ASAT system to counter the growing threat.[11] First-strike stability in space was beginning to decouple from nuclear crisis stability. Nevertheless, President Carter continued to seek arms-control agreements with Moscow, including a ban on ASAT weapons, but he also kept the U.S. ASAT program alive as a bargaining chip in those efforts and as a hedge against their failure. To the administration's disappointment, Washington and Moscow were unable to come to terms on an ASAT treaty in three rounds of negotiations between June 1978 and June

---

[9]   R. Cargill Hall, *Military Space and National Policy: Record and Interpretation*, Washington, D.C.: George C. Marshall Institute, 2006, p. 8.

[10]   For a historical review and eloquent defense of these arguments, see Ziegler, 1999, pp. 185–245. For a late–Cold War critique of what is sometimes called the sanctuary doctrine, see David E. Lupton, *On Space Warfare: A Space Power Doctrine*, Maxwell AFB, Ala.: Air University Press, September 1989.

[11]   Stares, 1985, p. 179; also see National Security Decision Memorandum 345, "U.S. Anti-Satellite Capabilities," January 18, 1977.

1979. A fourth set of meetings, expected to take place in the autumn of 1979, was delayed due to the U.S. debate over ratification of the SALT II agreement and then cancelled after the December 1979 Soviet invasion of Afghanistan.[12]

That event drove in the final coffin nail on détente, and, with Ronald Reagan taking office soon afterward, U.S.-Soviet relations returned to a confrontational tenor reminiscent of earlier periods. The Reagan administration immediately increased funding for ASAT development, and the program soon bore fruit. On September 13, 1985, the Air Force conducted a successful ASAT test, launching from an F-15 a modified short-range attack missile with an Altair III second stage that flew in direct ascent, destroying a target satellite via kinetic impact.[13] Now, with both superpowers experimenting with ASAT systems and finding ever more ways to support their conventional military operations from space, it was becoming increasingly clear that space had become a potential flashpoint of conflict. No longer the near-exclusive domain of national intelligence, surveillance, and nuclear command-and-control missions, space-based assets were providing increasingly valuable services to conventional military forces. This reality increased the likelihood that those assets might become targets of enemy attack to deny U.S. forces the advantages provided by space support.

The United States' orbital infrastructure had also become progressively more important to the nation in ways not directly related to national defense. An ever-greater volume of civil and commercial activities had come to rely on support from communication and meteorological satellites, and new markets were emerging in such areas as space-based spectral imaging for resource management and geodesic survey. Given this importance and the history of Soviet ASAT activities, administration officials and military leaders doubted that the United States would be able to rely on any tacit understanding preserving space as a sanctuary in a serious conflict with the Soviet Union. They concluded that space would likely be a battleground, and the

---

[12] Stares, 1985, pp. 180–200.

[13] Peebles, 1997, pp. 66–68.

United States needed to prepare accordingly.[14] Consequently, U.S. military leaders began both worrying about how they might deter or defend against attacks on U.S. systems and considering whether they could deny the use of space to adversaries.

Targets of Growing Attractiveness

The end of the Cold War muted such concerns for a while, but they reemerged as the United States repeatedly demonstrated its space-enabled dominance in conventional warfare. The 1991 Gulf War is often described as the "first space war" due to the many ways that space services were used in support of U.S. and coalition forces. But the space support provided in that conflict was only a foretaste of what was to come. In July 1995, the Global Positioning System (GPS) achieved full operational capability, with 24 satellites on orbit providing continuous, precise positioning, navigation, and timing (PNT) support to military and civilian users around the globe. With that capability came a whole new class of precision weapons—from gravity bombs to cruise missiles—using GPS data to guide them to their targets. Moreover, as advanced, space-enabled command-and-control systems were developed to integrate near-real-time intelligence, surveillance, and reconnaissance (ISR) and GPS data, a new generation of network-centric warfare concepts emerged, propelling U.S. forces toward a transformation in conventional warfighting effectiveness, which the United States repeatedly demonstrated in the post–Cold War era in conflicts from the Balkans to the Middle East to South Asia.

While such dramatic increases in capability have pleased U.S. leaders, they have also called attention to how much U.S. military forces have come to depend on space support. Many strategic thinkers

---

[14] Much of that preparation involved reorganizing military space operations to better integrate with and support conventional warfighting functions. The Air Force created Air Force Space Command for that purpose on June 21, 1982, and the Navy followed suit with Navy Space Command on October 1, 1983. The nation's first unified command for military space operations, U.S. Space Command, was inaugurated on September 23, 1985. The Army activated its service component, the Army Space Agency, in 1987, then reorganized it to form Army Space Command in April 1988.

have questioned whether, in any serious confrontation, an adversary capable of attacking U.S. space systems would refrain from doing so and thereby allow the United States to retain its conventional war-fighting advantage unchallenged. Some have pointed to vulnerabilities on the ground. Indeed, satellite ground stations and other portions of the space-support ground infrastructure have long been susceptible to attack, but the degree of threat they face in limited conventional conflict is probably not very great.[15] There is relatively little payoff in attacking most elements of the ground infrastructure because multiple satellite control stations and ground processors provide redundant capabilities for commanding satellites and receiving and processing critical data streams. Moreover, most satellite constellations could operate for days or even for weeks without any ground support, although mission effectiveness and satellite state-of-health would degrade over time. Finally, most satellite ground stations will always be outside the contested area in any particular crisis or limited war. Attacking them would violate U.S. sovereignty or the sovereignty of friendly states, thereby incurring risks of escalation. All things considered, ground infrastructure attacks present immediate risks to the perpetrator while offering little probability of significant near-term impacts on U.S. space capability, so satellite ground stations would probably not be attractive targets in limited conventional conflicts.

The orbital infrastructure is a different story, however. Satellites are fragile pieces of equipment that move in predictable paths devoid of geographical cover, so they are vulnerable to attack and difficult to defend. They are very limited in their ability to maneuver and extremely susceptible to kinetic impact of any kind, even from objects of very small mass. Satellites can also be attacked by directed-energy weapons, their sensors can be obscured, and their links can be jammed. International norms condemning attacks on satellites emerged over the

---

[15] Conversely, nonstate adversaries, such as terrorists and insurgents, may present significant risks to U.S. space ground stations in some locations. However, those actors' attacks would likely be motivated more by the fact that lightly defended space assets present a convenient opportunity to inflict U.S. casualties than by any serious effort to degrade U.S. space capabilities, although the threat to space capabilities implied in such attacks would also get media attention that might advance their cause.

course of the Cold War, and kinetic attacks in space create debris that endangers every spacefaring nation's assets in similar orbits. These factors suggest that attacking satellites would result in the aggressor facing some degree of international censure, but analysts who argue that this imposes stability on the strategic environment may be overestimating the deterrent leverage of such prospective costs. Attacking uninhabited satellites does not harm people directly, and a state facing the prospect of taking very real human casualties and sacrificing important national interests at the hands of a space-empowered opponent may not be discouraged from attempting to avoid or reduce those losses by the prospect of international criticism. While some capabilities exist to defend against certain kinds of attacks on satellites and their communication links, all are limited against determined attackers and all entail additional expense, discouraging commercial and some military satellite owners from investing in them. Consequently, since space systems are so difficult to defend, an offensive advantage exists for states willing and able to attack them, and first-strike stability is at risk in any confrontation with such an adversary.

First-strike instability is made worse by limitations in space situational awareness (SSA). While the United States enjoys better SSA than any other spacefaring nation, it is still dangerously limited. Not all satellites are monitored constantly, and only limited diagnostic and environmental monitoring capabilities exist even for those that are, making it difficult to diagnose causes of sudden satellite failure. Knowing this, adversaries might be tempted to attack satellites covertly, believing that uncertainty regarding the causes of failure would impede retribution, or perhaps even that attacks would be misdiagnosed as naturally occurring failures. Alternatively, a natural failure that occurs during a confrontation or conflict could lead operators or policymakers to assume that the satellite was attacked, prompting unjustified retribution and subsequent escalation of the crisis.

This dangerous combination of continued vulnerability, growing dependence, and limited SSA indicate that first-strike stability in space has diminished, and further indications suggest that the rate of erosion is accelerating. While the difficulty of attacking orbital assets remains a stabilizing factor, that factor is shrinking as an increasing number of states acquire capabilities to interrupt space services. Several states are now attempting to develop directed-energy weapons. One of them, Russia, also retains the co-orbital ASAT capability that the Soviet Union developed during the Cold War and has since sold GPS jammers to anyone with the funds to purchase them. As has been the case since the dawn of the space age, any state with ballistic missiles and nuclear weapons has the basic components to field a crude but highly destructive ASAT weapon.[16] The proliferation of such threats is troubling, and anxieties have become more acute now that China has begun experimenting with directed-energy weapons and has demonstrated a capability to destroy satellites in low earth orbit (LEO) with a direct-ascent kinetic ASAT weapon.[17] Unfortunately, the infrastructure, policies, and attitudes that both enable and constrain U.S. space operations in the current environment are, in many ways, unchanged from when they were developed during the MAD-induced stability of the Cold War. This leaves the United States exposed to the risk of a surprise attack in space unless a deterrence regime can be developed to restore first-strike stability in that domain.[18]

Estimated Thresholds of Space Deterrence Failure

While the foregoing analysis suggests that the level of first-strike stability that the United States and its potential adversaries have enjoyed

---

[16] For a comprehensive unclassified analysis of potential threats to U.S. space capabilities as they existed in 2000, see Tom Wilson, "Threats to United States Space Capabilities," prepared for the Commission to Assess United States National Security Space Management and Organization, 2000.

[17] U.S. Department of Defense, *Annual Report to Congress: Military Power of the People's Republic of China, 2007*, Washington, D.C.: Office of the Secretary of Defense, 2007.

[18] Concerns about U.S. vulnerabilities to a surprise attack in space were expressed more graphically in the 2001 final report of the Rumsfeld "Space Commission," which said that the United States is an attractive candidate for a "space Pearl Harbor." See Commission to Assess United States National Security Space Management and Organization, *Report of the Commission to Assess United States National Security Space Management and Organization*, submitted to the House Armed Services Committee, Washington, D.C., January 11, 2001, pp. xiii, xv, 22, 25.

to date is diminishing, it would be overly simplistic to assume that the thresholds of deterrence failure are the same for all space systems or at all levels of war. In any given crisis or conflict, there is a range of factors that an adversary would have to weigh in contemplating attacks on U.S. space capabilities, and the risks and potential benefits in a space attack would vary greatly in the context of specific scenarios. Consequently, it is less a question of whether would-be aggressors can be deterred from attacking U.S. space systems than of what kind of attacks against which capabilities could be deterred under what circumstances.

Different attacks bear different risks of retribution. Reversible-effects attacks, such as dazzling and jamming, that do not damage space system components would credibly justify much lower levels of punishment than would attacks that do cause damage. Attacks that physically damage a satellite increase the probability that the victim would attempt to punish the attacker in some costly way, but those that do not generate debris would probably not incur the same level of wrath as kinetic strikes that litter the space environment. Courses of action that prospective aggressors would likely consider even more risky include attacks that cause indiscriminate damage to the global orbital infrastructure, such as wide-scale use of space mines, or those that directly take human life, such as physical raids on ground stations before the onset of terrestrial hostilities. The economic implications of a major kinetic attack on space assets could be substantial. Given that most states with capabilities to pose serious threats to U.S. space assets are themselves heavily invested in the international financial system, they would have to take such implications into consideration. Among the most grievous space attacks would be the detonation of one or more nuclear devices in space—an act that would likely, over time, cause catastrophic damage to the orbital assets of all spacefaring nations.

Against this range of risks, the prospective space aggressor would have to weigh the degree to which the benefits of a particular option would serve a given objective in peace or at any specific point in a crisis or war. As Figure 2.1 illustrates, this assessment of costs and benefits would likely vary from one prospective space target set to another at each level of conflict, and the threshold of deterrence failure would be different for nondestructive attacks (i.e., reversible-effects attacks) than for destructive attacks (those that cause damage).

Some of those thresholds are quite low today. An opponent in a confrontation with the United States that has not yet engaged in conventional terrestrial hostilities might consider reversible-effects attacks on U.S. space-based ISR and communication assets to be a promising means of degrading the United States' ability to respond to the crisis, with a relatively low risk of serious retribution compared to that of a destructive attack on one or more U.S. satellites. Fearing the onset of U.S. air strikes, the adversary might also begin jamming GPS signals in areas around command-and-control nodes and other important facilities to degrade the accuracy of U.S. precision-guided weapons.

Once fighting has begun, space deterrence would become a function of escalation management. A savvy adversary might continue to abstain from destroying U.S. satellites in a limited war for fear of escalating the conflict, particularly if the reversible-effects attacks continued to yield comparable levels of benefit. Similarly, in a war limited in scope and time, the enemy would not be likely to attack space-support ground infrastructure outside the area of conflict, although ground stations inside the combat zone would likely be regarded as fair game if they could be reached.

However, should the terrestrial conflict escalate, it would become increasingly difficult to deter an enemy with the appropriate capabilities from carrying out destructive attacks in space. Threats to escalate the conflict by punishing the enemy in the terrestrial domain lose potency in proportion to the extent that such escalation has already occurred and such costs are already being paid. Furthermore, as a war intensifies, the number of U.S. and allied space assets on the enemy's target list would likely grow, potentially saturating the reversible-effects weapons that would be available. At some point, the conflict would reach a threshold at which the growing benefits of transitioning to destructive attacks on certain space systems would overtake the dwindling costs of doing so, and the enemy would escalate in space. Once again, a rational enemy would likely prefer to use means of destructive attack that would avoid creating debris, thereby risking less condemnation from the international community. However, other factors may intervene in this decision: By then, such weapons may have been destroyed, or the adversary may wish to withhold from using them to avoid their destruction; the conflict may have generated use-or-lose pressures on kinetic ASAT launchers; or, depending on the international political climate, the adversary might conclude that global censure for endangering the worldwide orbital infrastructure would fall more heavily on U.S. shoulders than on its own.

In any event, the adversary would become increasingly inclined to attempt destructive attacks on U.S. orbital assets as the conflict escalates. As Figure 2.1 indicates, some ISR satellites, such as those used for reconnaissance and ocean surveillance—being high-value, low-density assets—might become targets even at relatively low levels of conflict, and the adversary might attempt to damage dedicated U.S. military satellite communication (MILSATCOM) assets as well.

Conversely, since commercial satellite communication (SATCOM) platforms typically support a host of international users as well as U.S.

Figure 2.1
Notional Space Deterrence Capabilities, by System Type at Various Levels of Conflict

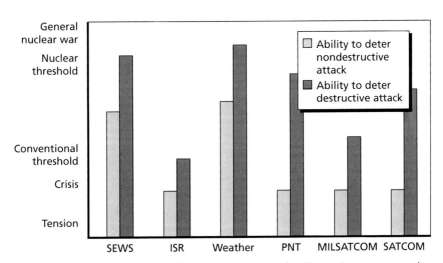

NOTE: The information shown in this figure is provided for illlustrative purposes only and is not based on an analysis of empirical data.
RAND *MG916-2.1*

forces, the political costs and escalatory risks of mounting destructive attacks on those assets would likely deter the opponent from attempting to do so until the conflict escalated to a higher level. Satellites supplying PNT data—i.e., GPS—would probably be relatively safe from destructive attack until very high levels of conflict, because the distributed nature of that system would make it difficult for an opponent to realize much benefit from individual attacks.

The adversary would also likely be deterred from damaging U.S. SEWS assets to avoid risking inadvertent escalation to the nuclear threshold, but that firebreak would almost certainly collapse with the conclusion that such escalation is inevitable and that it is in the adversary's interest to launch a preemptive nuclear strike. Moreover, because the strategic surveillance and warning system also supports efforts to locate and destroy mobile conventional missile launchers, the adversary might risk dazzling SEWS satellites at lower levels of conflict to impede U.S. efforts to locate and destroy those launchers.

Weather satellites, surprisingly, might be the space assets that are safest from attack. Attacking assets supporting the highly globalized international meteorological system would result in considerable political costs, and the robust infrastructure supporting that system would limit the benefits of individual attacks against it.

The longer a conventional war between the United States and an enemy capable of attacking space assets, the greater the pressure would be for escalation in both the terrestrial environment and space. Should the conflict expand in scope and expected duration in ways similar to major wars in the 20th century, the benefits of attacking satellite ground stations and other elements of the U.S. space-support ground infrastructure would grow and the prospective costs of doing so would shrink.

Alternatively, even if the war were to remain confined in duration and geographic scope but escalate to the point at which the enemy felt threatened by prospects of regime change—and especially if that were the United States' stated objective—then it would not be reasonable to expect that the United States could deter the enemy from resorting to any level of destructive attack in space, including the use of nuclear weapons, if it appeared that such actions might reduce the enemy's chances of defeat. While one might question why an adversary would expend a valuable and probably limited resource on what might seem to be a senseless act of destruction, exploding one or more nuclear weapons in space while keeping others in reserve to hold regional terrestrial targets at risk could, in some circumstances, be a rational and plausible tactic.[19] As serious as such an act would be, it would, without directly taking human life, effectively signal that severe levels of escalation in the terrestrial domains were imminent if the United States did not desist in pressing the offensive. Such an act would doubtless make the perpetrator a pariah in the international community, but many world leaders would consider that outcome preferable to losing their regimes and, potentially, their lives.

Space Deterrence and General Deterrence

Although this assessment focuses specifically on deterrence and first-strike stability in space, it is important to appreciate the interdependencies between these factors and general deterrence and stability writ large. Given the extent to which space support enhances U.S. conventional military capabilities, an adversary weighing the risks and potential benefits of war with the United States might be encouraged toward aggression by the belief that attacking space systems would degrade U.S. warfighting capabilities enough to enable the attainment of objectives at acceptable costs. As a result, weaknesses in space deterrence can undermine general deterrence. Conversely, if a prospective adversary concludes that the probable cost-benefit outcome of attacking U.S. space systems is unacceptable, it is forced to weigh the risks and benefits of aggressive designs in the terrestrial domain against the prospects of facing fully capable, space-enhanced U.S. military forces. In sum, effective space deterrence fortifies general deterrence and stability.

---

[19] For more on the escalation dynamics that could emerge in such confrontations with regional nuclear powers, see Morgan et al., 2008, pp. 83–115.

CHAPTER THREE

Applying the Principles of Deterrence to the Space Environment

Deterrence was the central pillar of U.S. strategic thought from the late 1940s until the end of the Cold War.[1] Yet, the fundamental mechanisms of deterrence are not unique to the Cold War or even the nuclear era. Throughout history, states have sought to deter potential enemies from attacking them by building strong defenses and powerful armies. But in the early 20th century, when the emergence of airpower provided a means to bypass enemy defenses and inflict pain on adversaries deep in their homelands, the dynamics of deterrence began to shift from erecting visible defenses to making threats of punishment. States built bombers, in part, to deter enemies from razing their cities by emphasizing the ability to inflict punitive costs in kind.[2] Later, the advent of nuclear weapons shifted the focus of deterrence to threats of punishment entirely, because no defenses could be devised that

---

[1] Literally hundreds of books and journal articles on deterrence were written over the course of the Cold War. Some of the most prominent and influential monographs include Bernard Brodie, ed., *The Absolute Weapon: Atomic Power and World Order*, New York: Harcourt Brace, 1946; Bernard Brodie, *Strategy in the Missile Age*, Santa Monica, Calif.: RAND Corporation, 1959; Thomas C. Schelling, *The Strategy of Conflict*, Cambridge, Mass.: Harvard University Press, 1960; Thomas C. Schelling, *Arms and Influence*, New Haven, Conn.: Yale University Press, 1966; Alexander L. George and Richard Smoke, *Deterrence in American Foreign Policy: Theory and Practice*, New York: Columbia University Press, 1974; and Richard Smoke, *National Security and the Nuclear Dilemma: An Introduction to the American Experience*, New York: Random House, 1984.

[2] For an analysis of how early-20th-century theories about using airpower to deter attack by threatening punishment influenced Cold War–era strategic thought on nuclear deterrence, see George H. Quester, *Deterrence Before Hiroshima: The Airpower Background of Modern Strategy*, New Brunswick, N.J.: Transaction Books, 1986.

were sufficiently reliable to save a state from the unacceptable costs of nuclear war.

Because space operations were so closely tied to national strategic missions during the first years of the space age, the emphasis on punishment-based nuclear deterrence that undergirded MAD and imposed stability in the terrestrial domain lent stability to space as well. But as first-strike stability in space became decoupled from nuclear crisis stability, the dynamics of deterrence in space changed. Employing space services in the support of conventional military operations puts space systems at risk in ways that are irrelevant to threats of nuclear retribution, because such threats are not credible in response to attacks on satellites not solely dedicated to nuclear missions. As this chapter explains, threats of punishment using conventional force, conversely, also may not be sufficiently potent or credible to deter attacks on space systems. Therefore, deterring attacks on these assets will require the ability to fashion capabilities more akin to those used for conventional deterrence in the terrestrial environment—those that persuade a prospective attacker that it cannot expect to reap sufficient benefit from attacking space systems to justify the likely costs. Unfortunately, it may be difficult to field such capabilities. To illustrate how hard reestablishing first-strike stability in space may be, let us review the fundamental principles of deterrence and explore the complex dynamics that emerge when those principles are applied in the space domain.

The Central Mechanism of Deterrence

Deterrence can be described as discouraging an adversary from doing something it might otherwise choose to do by manipulating its calculations of cost and benefit.[3] This approach involves persuading an opponent that if it were to take some prohibited action, one would inflict costs, deny success, or some combination of the two, so that the adversary concludes that the *probable costs* would outweigh the *probable ben-*

---

[3] As Alexander George and Richard Smoke maintain, "[I]n its most general form, deterrence is simply the persuasion of one's opponent that the costs and/or risks of a given course of action he might take outweigh its benefits" (George and Smoke, 1974, p. 11).

efits and therefore decides against acting. Deterrence is a form of coercive persuasion: Strategies designed to deter an opponent are intended to influence the behavior of a voluntary agent—one that retains the power to either abide by the threatener's demands or defy them. Strategies designed to deny the opponent choice in the matter are not deterrence; they are controlling strategies.[4]

While the foregoing description is relatively straightforward, it is important to appreciate the subjective, speculative nature of deterrence, because it is in these qualities that thinking about deterrence and fashioning strategies to achieve it become challenging. First, as two opponents are posturing to influence the outcomes of future events, the one being threatened cannot know with certainty what costs the threatener can impose or how successful its own attacks might be should it defy those threats. Thus, uncertainty and probabilities play an important role in the decision calculus.[5] The enemy must estimate risks, both the probable costs and the probability of not achieving sufficient success to make those costs worthwhile. But just as important is the subjective nature of deterrence. While most Cold War–era thinking about deterrence assumed that both antagonists would be rational actors weighing prospective costs and benefits on some objective scale, assessments of cost and benefit are, in fact, subjective—*and it is the enemy's assessment that counts, not the threatener's.*[6] Therefore, for deterrence to be effective, the threat must be sufficiently potent to reliably manipulate the opponent's decision calculations in the desired manner, and the

---

[4] Lawrence Freedman, *Deterrence*, Cambridge, UK: Polity Press, 2004, p. 26. Also see David E. Johnson, Karl P. Mueller, and William H. Taft V, *Conventional Coercion Across the Spectrum of Operations: The Utility of U.S. Military Forces in the Emerging Security Environment*, Santa Monica, Calif.: RAND Corporation, MR-1494-A, 2002, pp. 8–9. Thomas Schelling (1966, pp. 2–6) differentiated deterrence from controlling strategies as a contrast between coercion and brute force.

[5] John Sheldon adds that deterrence is an uncertain strategy for the deterrer as well. Therefore, no national security policy should ever rest on deterrence alone. (John B. Sheldon, *Space Power and Deterrence: Are We Serious?* Washington, D.C.: George C. Marshall Institute, November 2008, p. 2.)

[6] George and Smoke, 1974, pp. 54, 73–76; Schelling, 1966, pp. 229–232. Also see Richard K. Betts, *Nuclear Blackmail and Nuclear Balance*, Washington, D.C.: Brookings Institution Press, 1987, pp. 133–134.

threatener must convince the opponent that it has both the capability and the resolve to carry out the threat if the prohibited action is taken. Put simply, the threat must be both powerful and credible. Just as important, however, and often forgotten, is that in order to move the opponent's decision in the desired direction, there must be credible assurance, expressed at least implicitly, that the opponent will not be punished if it does not take the prohibited action. The opponent must be led to conclude that good behavior will leave it better off.

Efforts to apply this logical framework to the task of reestablishing first-strike stability in space reveal the challenges involved. Deterring an adversary from attacking U.S. space systems (satellites, ground infrastructure, and communication links) would require the United States to issue potent and credible threats of punishment, denial, or some combination of both. Employing the first approach would entail threatening sufficient punishment to persuade the opponent that the costs that would be suffered in response to attacks on U.S. space systems would likely outweigh any benefits achieved and that it would not pay those high costs if it withheld such attacks. The second approach would entail persuading the opponent that it cannot expect sufficient benefit from prospective attacks to make them worth the probable cost. Both approaches are logically viable, but making them sufficiently potent and credible to be effective will be difficult.

The Limits of Punishment-Based Deterrence in Space

Threats of punishment for attacks on space systems face unique challenges in terms of potency and credibility. The punishment-based approach that most readily comes to mind for deterring attacks on U.S. satellites entails threats of retribution against the opponent's satellites—the old "if you shoot ours, we'll shoot yours" model. Such a threat sounds reasonable and balanced; however, given the disproportionate degree to which U.S. forces depend on space support as compared to potential adversaries, it would probably lack sufficient potency to deter a serious opponent. Future enemies of the United States will probably be fighting in their own neighborhoods and employing operational concepts that rely less on space-based ISR and communication assets than do U.S. forces, so enemy leaders might even welcome a game of satellite tit-for-tat, as the benefits of denying space support to U.S. forces would likely outweigh the costs of losing their own assets in return.[7]

That said, the United States has greater strategic depth in space than any of its potential adversaries. The fact that it has more powerful conventional forces with warfighting capabilities enhanced by support from an orbital infrastructure that is much more developed than that of any other nation presents a formidable obstacle to any prospective challenger. However, it is important to remember that whether deterrence maintains or fails is more than a simple binary function. As explained in Chapter Two, an adversary in confrontation with the United States might well begin with nondestructive attacks—those that do not justify a costly punitive response—to degrade U.S. abilities to deploy and intervene in the region. But as the crisis intensifies, cascading events could escalate to the point that a conflict appears imminent and the opponent considers conducting destructive attacks on selected high-value, low-density orbital assets. Were the opponent to conclude that such attacks would increase its chances of military success, threatening to attack its satellites in return might have little deterrent effect.

Consequently, some analysts have suggested that the United States should threaten to punish space aggressors with conventional military attacks in the terrestrial environment. Indeed, the United States has substantial capability, mainly through the use of conventional airpower, to punish other international actors and has done so

_____

[7] Some observers have noted that, as other states invest more in space capabilities, the balance of orbital assets at risk will begin to shift, thus making prospective outcomes of satellite tit-for-tat more even and space deterrence more viable for the United States. That argument is valid to an extent. However, no state is likely to approach U.S. levels of investment in space in the foreseeable future. More importantly, given the facts that (1) U.S. military interventions almost invariably entail expeditionary operations against adversaries in their home regions and (2) the transformational warfighting capabilities that U.S. forces employ to maintain qualitative advantages (potentially in the face of quantitative disadvantages) are greatly enhanced by space support, U.S. military operations are likely to remain more dependent on space than those of other countries, even if levels of adversary investment in space do approach those of the United States.

on numerous occasions in the past. Yet, powerful as U.S. capabilities are, it may be difficult to make conventional threats potent enough to deter aggression against space systems when opponents face the prospect of war with the United States. Conventional forces, no matter how powerful, generally cannot inflict great costs on an adversary in a short period. Given sufficient time, conventional forces can impose terrible costs—indeed, they can eventually inflict costs comparable to those of limited nuclear attacks—but, contrary to the case of nuclear deterrence, would-be aggressors may anticipate that, if conventional punishment is unleashed, they will have some amount to time to test their ability to defeat it or at least weather the storm. With that in mind, an adversary considering an attack on U.S. space systems in the face of a threat of conventional punishment would weigh the prospective benefits of such an attack against the ability to defend against the expected punishment, how long the punishment could be endured, and whether the punishment would end before its accumulated costs exceeded the expected benefits from the attack.

Unfortunately, aggressive leaders tend to be risk-acceptant optimists. Experience suggests that deterring aggression in the terrestrial environment without nuclear threats generally requires persuading the adversary that the prohibited action would entail a substantial risk of defeat or, at least, a high risk of bogging down in a costly war of attrition.[8] Attempts to deter aggression in space by threats of conventional punishment in the terrestrial environment would lead the would-be aggressor to similar considerations, but the outcome of the decision would be skewed by an expectation that attacks, if successful, would likely reduce the United States' ability to impose retributive costs. Moreover, if locked in a confrontation with the United States, were the aggressor to conclude that war was inevitable, it would also realize that *it would eventually have to pay a higher cost if it did not attack U.S. space systems.* Damage limitation, therefore, becomes an important part of

---

[8] John J. Mearsheimer, *Conventional Deterrence*, Ithaca, N.Y.: Cornell University Press, 1983, pp. 23–24, 28–30. Also see Karen Ruth Adams, "Attack and Conquer? International Anarchy and the Offense-Defense-Deterrence Balance," *International Security*, Vol. 28, No. 3, Winter 2003–2004, pp. 45–83.

the adversary's calculation, potentially tipping the scales toward a decision to launch a preemptive first strike in space.[9]

Moreover, while threats of conventional punishment would need to be powerful to deter attacks on U.S. space systems, efforts to make them so could ultimately undermine their credibility. For a nation known to value its self-image and international reputation to issue threats that are credible, the threats must appear justified—that is, they need to be reasonably proportionate to the seriousness of the misbehavior—otherwise, the opponent might doubt the threatener's resolve to carry them out. For instance, if a reputable nation issued a threat to inflict carnage on enemy civilians in retribution for some minor aggression, it might not be believed: To carry out such punishment would result in serious moral and political costs for the threatener. Credibility may be further weakened when there is no clear, logical relationship between the misbehavior and the threatened punishment. A threat to bomb an adversary's port for occupying a disputed territory that is landlocked might lack sufficient linkage to be taken seriously.

Putting these considerations in the context of space, in a confrontation before the onset of war, threats to bomb targets in an adversary's capital or other major cities in response to a destructive attack on a U.S. satellite might be doubted, given the dubious linkage, escalation risks, and probable casualties and collateral damage that such a response would entail. Carrying out such a threat would require applying force in a highly escalatory manner that, depending on the broader geopolitical circumstances,[10] might be condemned in domestic and world opinion, despite the fact that the adversary would have techni-

---

[9] This dynamic is similar to that seen in crisis-stability calculations between nuclear-armed opponents in the terrestrial environment. The greater the perceived inevitability of war, the greater the pressure to launch a damage-limiting first strike.

[10] Such a scenario cannot be properly evaluated in abstract without considering the state that the United States is confronting, that state's power and prestige in the international community, and a wide range of other political, strategic, and operational factors. For instance, while the United States did not hesitate to open the first Gulf War with an intense air attack on Baghdad, if U.S. leaders threatened to bomb Beijing in response to an attack on a U.S. satellite during a confrontation over Taiwan or issued similar threats against Moscow or St. Petersburg during a confrontation over Ukraine, they may not be believed, given those states' stature in the international community and their ability to respond to such attacks

cally crossed the threshold of hostilities first by launching an attack in space that destroyed one or more satellites. That attack would not have taken human life directly, nor would it have been easily observable to third parties. Weighing these considerations, the adversary might well conclude that such a threat is a bluff and risk attacking orbital assets. Threats to respond with punitive strikes against ASAT launchers, ground-based directed-energy weapons, or other portions of the adversary's counterspace architecture, such as tracking systems or command-and-control nodes, would have better linkage in that they are more clearly relatable to the act to be deterred. However, these threats might also be doubted in many scenarios because carrying them out would likely result in horizontal escalation. Such targets are likely to be outside the area in which the limited conflict is being fought. Striking them would broaden the scope of the conflict, inviting the enemy to respond with its own attacks on targets outside the area of operations. Even if believed, the threats might lack potency, given the resiliency of dispersed networks and the difficulty of finding and destroying mobile weapon systems.[11] Moreover, the adversary might not attach a high cost to the prospective loss of ASAT infrastructure if it believed that it could inflict severe and irreparable damage on U.S. space assets before effective counterstrikes could be carried out. Threats made in efforts to deter reversible-effects attacks before the onset of lethal hostilities suffer even more from defects in linkage and proportionality, and those made after combat has ensued would be largely irrelevant from a deterrence perspective.

The Difficulties of Denial-Based Deterrence in Space

Efforts to deter would-be aggressors by persuading them that the United States can deny them the benefits of attacking its space capa-

---

with counterstrikes of their own that would further escalate the conflict, potentially to catastrophic levels.

[11] One can even envision a strategy in which, during a crisis, an enemy attacks a U.S. satellite to deliberately provoke a punitive strike on its homeland, providing a very visible casus belli for war.

bilities also face serious challenges. While the United States should always emphasize the resilience of its space systems in order to discourage potential adversaries from attacking them, several factors may make this difficult. First, it is necessary to assume that potential adversaries are well aware that the transformational capabilities that give U.S. military forces their qualitative advantage are significantly enhanced by space support. They are likely to believe that attacking U.S. space systems offers a high payoff, because even limited success in attacks on some high-value, low-density assets might provide substantial warfighting benefits. Second, future enemies will also understand how difficult it is to defend space assets. Satellites possess inherent vulnerabilities, and all claims to the contrary are unlikely to be believed until proven. That presents a problem. There are passive defenses that the United States can employ to make satellites somewhat more resilient, but unlike visible forces and fortifications in the terrestrial environment, passive defenses on satellites are not observable in ways that deter attack. Electromagnetic pulse (EMP) shielding, radio frequency (RF) filters, and shuttered optics are not visible to the naked eye or even observable in the data collected by space surveillance systems. In fact, some defenses may need to be concealed in order to remain viable, thus eliminating the deterrent value of their existence. Consequently, the challenge will be to find ways to reduce the prospective benefits of attacking U.S. space systems that are demonstrable to potential enemies without undermining their effectiveness. Several approaches are possible, but all of them suffer certain limitations.

One option is to explore the extent to which the United States can develop more active ways to defend satellites via such capabilities as enhanced maneuverability or onboard active defenses. Enhancing satellite maneuverability for defensive purposes would require improving propulsion systems on satellites so that they could evade vehicles that attempt to intercept and destroy them. However, the extent to which enhanced maneuverability is possible is constrained by the "tyranny of orbital mechanics." It takes a great deal of energy to make any substantial change in the direction of movement of an object following Kepler's laws of motion at orbital speeds (approximately 17,000 mi/h, or 7,600 m/s, in LEO). Today's satellites, once separated from the rocket boosters used to place them on station, can do little more than effect marginal changes in velocity (delta-V), because their maneuvering thrusters are designed only for orbit maintenance and attitude control.[12] Improvements to this capability for most satellites will probably be limited to increases to the rate of delta-V, versus substantial changes in altitude or orbital plane. Doing anything more would require adding a more powerful propulsion system to the orbital platform or keeping a rocket booster attached to it during the operational mission. Both of those approaches present technical challenges and would add mass and, therefore, cost to the satellite. Satellite owners would have to weigh these costs against the limited benefits that capabilities for defensive maneuver might offer. It would be difficult for even a maneuverable satellite to evade a direct-ascent ASAT system, given short warning, and co-orbital ASAT systems can be made smaller, less massive, and therefore more maneuverable with less fuel expenditure than most of the satellites they would be designed to target.[13]

Alternatively, one can envisage fitting out satellites with onboard active defenses, such as short-range kinetic or directed-energy weapons designed to disable or destroy other space vehicles that come into close proximity, or even developing escort satellites with such capabilities. But once again, these ideas, while attractive in principle, would all require technical advances beyond what is possible today, and they would add cost to each mission on which they are flown. Moreover, adding any onboard defenses, active or passive, would be a long-term

---

[12] The primary life-limiting factor of a satellite is fuel. Maneuvers to maintain a satellite in LEO typically require a total delta-V of about 100 m/s per year, which means that a satellite with a planned 15-year lifespan would have a total delta-V capability of only ~1,500 m/s. A one-degree orbital shift at LEO would require roughly a 140 m/s delta-V, and an 11-degree orbital shift would consume all the fuel of a brand-new satellite. Alternatively, the satellite could speed up or slow down by almost 19 percent while consuming all of its initial fuel supply. Consequently, today's satellites are generally designed for low-thrust maneuvering over a given period. In fact, because of the extremely low-thrust advanced electric propulsion systems in use on some satellites, those spacecraft would take weeks or months to execute large orbital shifts.

[13] Given these advantages, even if a co-orbital ASAT system failed to intercept its target on the first attempt, it would likely have additional intercept opportunities on subsequent orbital passes, forcing the target satellite to maneuver repeatedly, expending precious fuel.

solution at best, as they could not be retrofitted to platforms already in orbit; rather, they could be installed only on new satellites. Replacing operational satellites before they fail would almost certainly be unaffordable, so fully upgrading the existing orbital infrastructure would be at least a decade-long project even if the technology were available today.[14] Escort satellites also face significant affordability challenges, as each of them could defend only one satellite at a single altitude and orbital plane, due, once again, to the rigid constraints that orbital physics impose on satellite maneuver.[15] More seriously, for any active defense to be viable, whether evasive maneuver or defensive counterstrike, the defensive system would need the ability to detect an approaching threat, analyze the critical parameters of the attack, calculate the appropriate defensive response, and execute that response before the attack culminated in the destruction of the defended satellite. Given today's limitations in SSA, the United States does not even have an observe-orient-decide-act (OODA) circuit that is fully functional in real time, much less an OODA loop that can be made tight enough to overcome an attacker's first-move advantage.

Missteps That Might Further Reduce First-Strike Stability in Space

Given the importance of space systems to U.S. national security, some academics and security analysts have argued that the United States should "seize the high ground" and place counterspace weapons in orbit to impose space dominance in the event of a conflict with another spacefaring nation. While such arguments resonate with those acculturated in the U.S. military tradition, it is hard to conceive how placing counterspace weapons in orbit would do anything to defend U.S.

---

[14] However, assets that operate in very low orbits, such as reconnaissance satellites, have to be replaced more frequently. Given available technology and funding, that portion of the infrastructure could be upgraded much sooner.

[15] If multiple satellites are put in the same orbital plane and at the same altitude, it is conceivable that an escort could be redeployed from the defense of one to another, but it could escort only one at a time, and any redeployment would expend propellant, shortening its life.

satellites from enemy ground-based weapons or, for that matter, other weapons in space. Rather, given the inherent vulnerability of satellites, placing weapons in orbit would increase first-strike instability in space by threatening potential adversaries with weapons that cannot, themselves, be defended. Taking this step may also encourage other spacefaring nations to follow suit, ultimately resulting in a dangerously unstable strategic environment that would generate severe "use-or-lose" pressures in the event of a military confrontation, whether the crisis originated in space or the terrestrial domain. Terrestrial-based counterspace weapons also endanger first-strike stability, particularly if states that invest in them exhibit brandishing behaviors, publicizing intentions to use them at the onset of conflict. But pressures to use terrestrial-based weapons first would not be as great, because they would not be as vulnerable to enemy action as space-based weapons.

Ballistic missile defense (BMD) systems that also have ASAT capabilities would likely affect first-strike dynamics in space in ways that mirror counterspace weapons. Systems with orbital components that could attack other satellites would, in a crisis with another space-faring nation that also had ASAT capabilities, exert pressure on that state to strike first, in an effort to save its own satellites from first-strike losses.[16] Similarly, terrestrial-based BMD weapons capable of intercepting satellites, might also be threatening to a spacefaring opponent in a crisis, but first-strike pressures would not be as great as they would be if either of the adversaries had weapons in orbit.

In all of the foregoing cases, brandishing behaviors would make first-strike instability more severe, given space systems' inherent vulnerabilities, as might explicit deterrent threats if they are not carefully tailored to support a coherent national strategy to enhance first-strike stability in space.

---

[16] While serving as a staff officer in the Headquarters Air Force Directorate of Strategic Planning (AF/XPX) in the late 1990s, I saw these dynamics play out in several futuristic war games in which both sides were armed with orbital laser constellations fashioned as BMD weapons with counterspace capabilities. In every instance, although the scenario began with a crisis short of war, each team initiated hostilities in turn one, expending its lasers against its opponent's space weapons as quickly as possible, knowing that to hesitate would mean losing its own weapons before they could be used.

All of this suggests that it may be difficult to deter potential enemies from attacking certain U.S. space systems in some circumstances. Nevertheless, the task of strengthening first-strike stability in space is by no means impossible. As illustrated at the end of Chapter Two, the orbital infrastructures of some U.S. systems are already sufficiently robust that they present poor targets for prospective attackers. Conversely, other systems are both vulnerable and important to U.S. military operations, making them attractive targets for adversaries capable of attacking them and willing to pay the retributive costs of doing so. The challenge will be to find ways to raise the thresholds of deterrence failure for vulnerable systems by manipulating potential adversaries' calculations of expected costs and benefits of attacking them, simultaneously raising their fears of costly punishment while lowering their expectations of success or warfighting effect. Meeting this challenge will require the United States to develop and employ a coherent national space deterrence strategy.

CHAPTER FOUR

The Need for a National Space Deterrence Strategy

Strengthening first-strike stability in space will be challenging, given the inherent vulnerability of some space systems and the extent to which the United States depends on the services they provide. In fact, for those systems that are most vulnerable and militarily important, it may be that no *single* means of deterrence, whether based on threats of punishment or on measures taken to deny benefits of attack, will be sufficiently potent and credible to reliably discourage a capable adversary from attacking them if it appears that war is inevitable. Yet, even if no single measure holds much promise of affecting an opponent's decision calculus in the desired manner, it may still be possible to dull future enemies' ardor for space aggression by manipulating both sides of their cost-benefit calculations simultaneously. Such an approach would require a sophisticated, multifaceted strategy incorporating threats of punishment in several dimensions—diplomatic and economic, as well as military—while also employing multiple mechanisms to persuade potential opponents that attacking U.S. space systems will not yield them sufficient benefit to justify the inevitable costs of doing so. For such a strategy to be viable and coherent, it would need to be employed as part of a broader national space strategy developed using a top-down, strategy-to-tasks approach.[1]

---

[1] The strategy-to-tasks methodology is a framework for defense planning developed at RAND in the early 1980s by Glenn A. Kent. Since that time, the approach has evolved and been applied to a wide variety of planning challenges. A partial list of references includes Glenn A. Kent, *Concepts of Operations: A More Coherent Framework for Defense Planning*, Santa Monica, Calif.: RAND Corporation, N-2026-AF, 1983; Edward L. Warner III and

The essence of the strategy-to-tasks methodology is establishing relevant top-level policy objectives first and developing a comprehensive strategy for achieving those objectives, then teasing out the necessary tasks for employing that strategy, ensuring that all actions are coherent and consistent with higher policy objectives. Such an approach will be vital for establishing a national space strategy designed to, among other things, effectively deter potential adversaries from attacking vulnerable U.S. space capabilities. The nation cannot afford for segments of the U.S. defense and policymaking communities to do or say things that work against first-strike stability in space while other elements are attempting to deter attacks on U.S. space systems.

National Space Policy

The foundation and central pillar of a national space deterrence strategy should be a national space policy designed to reinforce already emerging international taboos against space warfare by explicitly condemning attacks on space systems. Although the United States, Soviet Union, and China have all experimented with capabilities to destroy satellites, no state has yet attempted a destructive attack on another state's orbital assets. Now, 50 years into the space age, every additional year that passes without an attack in space persuades more citizens, business interests, and governments around the world that space warfare can be avoided and, due to the negative effects it could have on the operating environment shared by all spacefaring nations, ought to be

---

Glenn A. Kent, *A Framework for Planning the Employment of Air Power in Theater War*, Santa Monica, Calif.: RAND Corporation, N-2038-AF, 1984; Glenn A. Kent, *A Framework for Defense Planning*, Santa Monica, Calif.: RAND Corporation, R-3721-AF/OSD, 1989; and Glenn A. Kent and William Simons, *A Framework for Enhancing Operational Capabilities*, Santa Monica, Calif.: RAND Corporation, R-4043-AF, 1991. For modifications of the approach for application in resource management planning, see Leslie Lewis, James A. Coggin, and C. Robert Roll, Jr., *The United States Special Operations Command Resource Management Process: An Application of the Strategy-to-Tasks Framework*, Santa Monica, Calif.: RAND Corporation, MR-445-A/SOCOM, 1994, and John Y. Schrader, Leslie Lewis, William Schwabe, C. Robert Roll, Jr., and Ralph Suarez, *USFK Strategy-to-Task Resource Management: A Framework for Resource Decisionmaking*, Santa Monica, Calif.: RAND Corporation, MR-654-USFK, 1996.

prohibited. The George W. Bush administration effectively marshaled those sentiments to focus international censure on Beijing for conducting its direct-ascent ASAT test in January 2007, but the overall U.S. stance on space warfare is not conducive to first-strike stability. Current national space policy, unclassified and publicly available, explicitly charges the Secretary of Defense with maintaining capabilities to execute space control and force-application missions and, if directed, deny freedom of action in space to adversaries.[2] Unclassified military doctrines and strategic plans express aspirations for carrying out such missions,[3] and defense budget documents reveal programs to develop capabilities to apply force to and from the space domain.[4] While efforts to develop such plans and capabilities may be prudent, openly expressing U.S. intentions to dominate space does nothing to deter others from attacking U.S. space systems; rather, given the first-strike advantage so prevalent in the space strategic environment, it animates the efforts of potential adversaries to develop similar capabilities and, in a crisis, would provide motive and justification for their preemptive employment.

A national space policy more conducive to deterring attacks on U.S. space systems would avoid provocative rhetoric about denying others the use of space and would, instead, explicitly condemn any use of force to, from, or in that domain, except in retribution for attacks on one's own space systems. The United States could continue research into technologies needed for degrading or destroying enemy space capabilities, but the explicit organizing rationale for such efforts should

---

[2]   See Office of Science and Technology Policy, Executive Office of the President, *U.S. National Space Policy*, Washington, D.C.: White House, August 31, 2006.

[3]   See U.S. Joint Chiefs of Staff, *Space Operations*, Joint Publication 3-14, Washington, D.C., January 6, 2009; U.S. Air Force, *Space Operations*, Air Force Doctrine Document 2-2, Washington, D.C., November 27, 2006; U.S. Air Force, *Counterspace Operations*, Air Force Doctrine Document 2-2.1, Washington, D.C., August 2, 2004; and Air Force Space Command, *Air Force Space Command Strategic Master Plan: FY06 and Beyond*, Peterson AFB, Colo., October 1, 2003.

[4]   See Office of the Under Secretary of Defense (Comptroller), *Department of Defense Budget Fiscal Year 2009, RTD&E Programs (R-1)*, Washington, D.C., February 2008, line numbers 40, 54, 66, and 188.

be deterrence or, failing that, space defense, versus space control. U.S. leaders might even consider declaring a no-first-use policy regarding U.S. counterspace capabilities, but further analysis would be needed to determine whether such a policy would be preferable to leaving potential adversaries uncertain of where U.S. thresholds lie.

Another area in which the United States should revise its national space policy is that regarding its stance on space arms control. Current national space policy states, "The United States will oppose the development of new legal regimes or other restrictions that seek to prohibit or limit U.S. access to or use of space."[5] Such a position all but closes the door on proposals for arms-control treaties, in that they would, by definition, limit the ways in which the United States could use space.

While it is debatable whether treaties prohibiting space weapons will ever be viable, given the dual-use nature of technologies that could be used for attacking space systems, it is unwise to refuse diplomatic engagement on the issue. Agreements to abstain from certain behaviors can build confidence between states, and it is arguable that treaty negotiations can be valuable in and of themselves, whether or not agreements are ever reached. Although the United States and Soviet Union did not reach agreement in three rounds of ASAT treaty negotiations in the late 1970s, both parties benefited from the experience in that each learned a great deal about the other's position on a wide range of issues. Moreover, even if comprehensive treaties are not possible, states can often hammer out useful agreements in selective areas. For instance, a treaty prohibiting the test of any weapon that creates orbital debris would be verifiable and beneficial to the entire spacefaring community. But more to the point, a national space policy that makes a priori statements refusing to consider arms-control agreements is not conducive to space stability in that it suggests to other actors that the United States may be pursuing space weapons for offensive purposes.[6]

Until any ASAT agreement is reached, however, the United States should continue research on terrestrially based capabilities for degrad-

---

[5]   Office of Science and Technology Policy, 2006, p. 2.

[6]   For a deeper examination of these issues, see MacDonald, 2008, pp. 27–31, and Harrison et al., 2009, pp. 19–20.

ing or destroying enemy space systems.[7] As a sovereign state, the United States has the responsibility to see to its own security, and U.S. leaders would be remiss if they did not strive to maintain the nation's technological advantage in such capabilities while others are free to develop them. But once again, the organizing rationale for such efforts should be deterrence and defense, not space dominance, per se. A deterrence-oriented national space policy would declare that the United States will severely punish any attacks on its space systems and those of friendly states in ways, times, and places of its choosing, thereby laying a foundation for other statements and actions designed to enhance the credibility of threats of punishment in both the terrestrial domain and space.

That said, in the event of war in the terrestrial domain, the possession of such capabilities would inevitably raise the question of whether the United States should employ them to advance U.S. military objectives. For instance, if the adversary is using a reconnaissance satellite to target U.S. ground forces or an ocean surveillance satellite to locate and target a U.S. carrier task force, should the United States use space weapons to neutralize that threat? While some analysts might promptly answer in the affirmative, the issue is not as straightforward as it might seem.[8] While doing so might save U.S. lives and deny the enemy a tactical advantage, decisionmakers would have to weigh those benefits against the potential costs of crossing the first-strike threshold in space. A host of questions would have to be considered. Could the threat be neutralized via reversible-effects attacks, or would the United State have to destroy some number of the adversary's satellites? Would the adversary have capabilities with which to retaliate against U.S. space systems? If so, what costs might be inflicted on U.S. space assets, how might those costs inhibit U.S. forces in the accomplishment of their military objectives, and what additional costs in "blood and treasure" might U.S. forces pay as a result? Were U.S. forces to damage or

---

[7] As explained later in this chapter, positioning weapons in orbit would be destabilizing and is, therefore, inadvisable.

[8] For arguments in favor of attacking enemy space systems in times of war, see, for instance, Gray and Sheldon, 2000, p. 245; Worden, 2000; and Sheldon, 2008, p. 4.

destroy enemy satellites, what political costs might the United States pay for being the first to take war into space? Finally, what impact would such a precedent have on international norms and the space deterrence regime, and what long-term costs might the United States pay as a result?

Some of these questions cannot be answered outside the context of an actual conflict. In some future scenario, U.S. leaders might well decide, after weighing the risks, benefits, and alternatives, to attack an enemy's orbital infrastructure. But in the meantime, U.S. leaders should be open to diplomatic engagement, treaty negotiations, and other confidence-building measures, and they should actively pursue agreements when they can be crafted to serve U.S. interests. In addition to the benefits that such agreements might offer, demonstrating leadership in diplomatic venues is important for characterizing the United States as a responsible world actor with the moral authority to use its power to protect the common operating environment of all spacefaring nations. In these and other settings, all U.S. policies, statements, and actions should be carefully orchestrated to foster and strengthen an international norm that condemns all but retributive attacks on space systems. Advancing such a norm would raise the political costs of space aggression in ways that potential adversaries would have to factor into their decision calculations in any crisis in which they are tempted to attack orbital assets.[9]

Deterring Attacks in Space with Threats of Punishment

Important as they are, norms alone will not deter aggression in space. When confrontation turns to crisis and it begins to appear that war is inevitable, the international political costs of violating peacetime norms of behavior pale in comparison to the costs of not taking action to reduce a dangerous adversary's warfighting capabilities. However,

---

[9]   Bruce MacDonald develops a similar argument for a national space deterrence strategy by sketching out three broad doctrinal options—dominance, deterrence, and arms control—then explaining how the second one, deterrence, best supports U.S. interests. See MacDonald, 2008, pp. 17–26.

fortifying taboos against attacking space assets would strengthen deterrence in another important way: It would bolster the credibility of U.S. threats to punish any state that violated the norm. As the space warfare taboo strengthens, U.S. policymakers could capitalize on leverage from it to generate support for diplomatic and economic sanctions against states that openly develop and test weapons for attacking satellites. More importantly, a firm stance condemning aggression in space, coupled with a national space policy that explicitly threatens those who attack space assets with severe punishment in ways, times, and places of the United States' choosing, would bolster the credibility of U.S. threats to strike targets in the terrestrial domain in retribution for attacks on U.S. space assets. The aim of U.S. declaratory policies and strategies should be to manage perceptions: The international community should be conditioned to accept the justice of punishing space aggressors in the terrestrial environment and support the United States in its use of lethal force to do so. Potential adversaries, in turn, should be conditioned to take seriously U.S. threats to strike terrestrial targets in exchange for attacks on its satellites. Granted, carrying out such threats could be highly escalatory in some scenarios, *but that is exactly the point.* If, by the consistent nature of U.S. policies and the explicit nature of U.S. statements, potential adversaries are convinced that the United States would inexorably carry out its threats regardless of the risks—indeed, were they led to believe that U.S. leaders had placed themselves in a position in which they could not do otherwise—the last clear chance to avoid catastrophic escalation is put squarely on the adversaries' shoulders. It places on them the onus of triggering a chain of events that might lead to a wider war.[10]

As previously stated, the United States should also continue research on capabilities for attacking enemy satellites. Although a simple tit-for-tat exchange of satellites would not work to U.S. strategic advantage, potential enemies must not be allowed to believe that

---

[10] This argument draws heavily on the concept of "brinkmanship," the strategy that Thomas Schelling proposed for managing escalation between nuclear-armed adversaries by manipulating the shared risk of war. For more on brinkmanship, see Schelling, 1966, pp. 91 and 99–125.

they could attack U.S. satellites without suffering costly losses to their own orbital assets in return. To make such deterrent threats credible, capabilities to carry them out would be needed, but until technological advances overcome the inherent vulnerability of satellites, all capabilities for attacking enemy space systems should be based in the terrestrial domain to better protect them and minimize first-strike instability in crises and war. To remain consistent with a national space policy as outlined here, the purpose of such systems would be to provide a credible deterrent threat of retribution and, failing that, viable capabilities for defending the nation's security interests in space. Any accusations that such capabilities are intended for dominating space or denying other states' access to that domain should rightly be dismissed as contrary to U.S. policy except when employed in response to an aggressor's first strike.

Enhancing Space Deterrence by Denying Adversaries the Benefits of Attack

Even a multifaceted, punishment-based deterrence strategy may not be sufficiently potent or credible to discourage an adversary facing the prospect of war with the United States. Therefore, a comprehensive U.S. space deterrence strategy should also focus efforts on persuading potential adversaries that the probability of obtaining sufficient benefit from attacking space assets would not be high enough to make it worth suffering the inevitable costs of U.S. retribution. Part of such a strategy would entail perception management: The United States should, to the greatest extent possible, conceal vulnerabilities of its space systems and demonstrate the ability to operate effectively without space support. However, perception management can only go so far in the face of observable weaknesses. Therefore, the strategy should also pursue multiple avenues to make vulnerable U.S. space systems more resilient and defendable, thereby demonstrating tangible capabilities to deny potential adversaries the benefits of attacking in space. An added benefit to the United States of incorporating such denial approaches in the national space deterrence regime is that they would make the services that space systems provide more robust against loss should deterrence fail.

Although satellites are inherently difficult to defend, those who design, procure, and operate space systems should, to the extent feasible and affordable, invest in capabilities to do so. Passive defenses—such as shuttered optics, shields and filters against EMP and RF attack, onboard subsystem redundancy, antijam technologies, and so on—should be installed on all future high-priority military and intelligence satellites, and the Department of Defense and Air Force should explore the possibilities of subsidizing some of these capabilities on commercial systems supporting national security missions. Efforts should be made to develop the necessary enhancements to propulsion systems and propellant capacities to improve satellite maneuverability, along with the ability to detect, assess, and respond to threats quickly enough to evade them. Research should continue in efforts to develop onboard active defenses, and novel approaches, such as microsatellite escorts, should be fully explored. Because passive defenses and many active defense systems are not readily observable, they contribute nothing to deterrence unless would-be attackers believe or, at least, suspect that they are in place. Consequently, as stated earlier, perception management will continue to be an important dimension in the tacit communication between U.S. authorities and prospective attackers. Specific vulnerabilities of U.S. space systems must never be divulged, and the resilience of the orbital infrastructure and its defenses should be emphasized wherever plausible. Nevertheless, it is necessary to acknowledge the fiscal and technical constraints on space defense, at least for the foreseeable future, and remember that potential adversaries are generally aware of those limitations as well.

Therefore, in addition to the foregoing efforts, the United States should strive to reduce the potential benefits of attacking its space systems by dispersing the capabilities they provide across a larger number of platforms and by placing redundant capabilities on orbit. Today, many national security space missions are hosted on platforms that support multiple payloads and users. Similarly, some missions, such as imagery collection, require satellites that are large, expensive, and easily detected and tracked. Both conditions have evolved for sound, practical reasons: The first is the result of efforts to manage the high costs of space lift with maximum economic efficiency, and the second is driven by mission requirements. Nevertheless, they both concentrate capabilities into nodes that are lucrative targets of attack, offering substantial payoffs to potential adversaries. A strategy to reduce the benefits of such attacks would be, to the extent feasible and affordable, to disperse missions onto separate platforms and place redundant capabilities on orbit. Ideally, new systems would be designed around distributed, multisatellite technologies, such as those used by GPS.[11]

An added benefit might be to distribute U.S. national security payloads across satellites owned by a range of other nations and business consortia friendly to the United States and also engage in data-sharing arrangements with them. Creating such "entanglements" would help deter would-be aggressors in several ways. Not only would it reduce the benefit gained in any single attack, it would also increase international support for the United States to punish attackers (thereby building additional credibility in threats to do so), and it would confront prospective attackers with serious risks of horizontal escalation in that attacking a shared "international security space infrastructure" might bring more states into the conflict.[12]

In sum, effective dispersal and redundancy would alter an adversary's cost-benefit calculation, making it less willing to suffer the costs of retribution for an attack that would provide only marginal benefits. However, there would be limitations to this approach. Some missions do not lend themselves to multisatellite solutions, and some nations or commercial operators may not want to host U.S. national security

---

[11] One possible approach is illustrated in the Defense Advanced Research Projects Agency's "System F6" concept. Also known as the "Future, Fast, Flexible, Fractionated, Free-Flying Spacecraft United by Information Exchange" program, its objective is to demonstrate the feasibility and benefits of a satellite architecture wherein the functionality of a traditional "monolithic" spacecraft is replaced by a cluster of wirelessly interconnected spacecraft modules. For more on this concept, see Defense Advanced Research Projects Agency, Tactical Technology Office, "System F6," Web page, undated.

[12] However, as John Sheldon points out, such an arrangement would work only if allies had equal access to data from U.S. satellite systems, something that would "require a substantial change in the secretive culture of the U.S. national security space community" (Sheldon, 2008, p. 3).

payloads on their platforms, thereby making them more attractive targets. The high costs of space lift may make dispersal and redundancy inordinately expensive, and for some missions, multiple small payloads may not provide sufficient capability to substitute for fewer satellites carrying large packages.

Another approach to reducing an adversary's benefits in attacking space systems would be to provide redundant capabilities using terrestrial backups. Indeed, such solutions are currently being pursued. Undersea cables and other terrestrial links already provide reach-back communication from well-established forward areas of operation, although they are vulnerable to sophisticated attackers (or accidents, such as the recent Mediterranean fiber cut). High-altitude lighter-than-air craft and long-endurance unmanned aircraft systems offer possibilities to supplement space-based platforms for some ISR and communication missions. The type of assets currently being developed would not be survivable in areas where an adversary could challenge friendly control of the airspace, but long-endurance aerial surveillance could, to some extent, supplement space capabilities on the periphery of an area of operations, and platforms flown in secure airspace could be used to relay some links inside the battlespace, thereby reducing the payoff an aggressor might yield in attacking satellites supporting parallel missions. Such options merit further exploration and development.

Concealment and deception tactics could also add to deterrence by reducing the benefits that a potential aggressor could expect to reap in an attack on U.S. satellites. While secret capabilities contribute nothing to deterrence, knowing that the United States has certain capabilities on orbit, but not being able to determine exactly where they are, may contribute a great deal. Even if an adversary thinks it can detect, identify, and track most critical U.S. space assets, concealment and deception tactics create uncertainty, complicating the targeting problem and thereby reducing the probability of success and expected benefit. Such tactics might work well in conjunction with the multi-nation dispersal approach mentioned earlier, as that would reduce the vulnerability of platforms hosting U.S. payloads.

Finally, the United States needs to continue efforts to make its space lift system, as well as its satellite manufacturing capabilities, more responsive in order to demonstrate U.S. capabilities for rapid replenishment. Faster replacement of lost satellites means a smaller tactical benefit for an opponent that attacks them. Because other means of deterrence by denial require technological advances and costly changes or augmentation to the existing orbital infrastructure, rapid replenishment and terrestrial backup are probably the best near-term avenues for denying the benefits of an attack on U.S. space assets.

The Critical Need for Better Space Situational Awareness

While many options exist for punishing space aggressors and reducing the benefits of their attacks, nearly all of them depend to some degree on improvements in SSA. Poor SSA undermines the credibility of threats of punishment in some scenarios, as the attacker may expect to have a reasonable chance of striking anonymously. Conversely, good SSA has intrinsic deterrent value, because any prospective aggressor, knowing that culpability for an attack might be quickly determined and exposed to the world, would have to weigh the long-term costs of angering the United States and international community, even if no immediate capability existed to inflict punishment. All active defenses require better SSA than what current capabilities provide, and many passive defenses could also be improved with better SSA. Lack of effective SSA could both inhibit the United States from taking reprisals against covert space aggressors and create risks that unjustified reprisals may be taken in response to natural satellite failures occurring during a crisis. Better SSA will improve diagnostic capabilities, helping operators to distinguish satellite malfunction from attack more quickly and reliably, thereby enhancing first-strike stability in space. Improving SSA should be one of the United States' top priorities in its efforts to develop the capabilities needed for an effective space deterrence regime.

Conclusion

Although strengthening first-strike stability in space will be challenging, the United States can do so if it develops a coherent national space deterrence strategy that operates on both sides of a potential adversary's decision calculus. Such a strategy would have to be sophisticated and multifaceted, incorporating threats of punishment in several dimensions while simultaneously pursuing multiple approaches to persuade potential opponents that attacking U.S. space systems would not yield them sufficient benefit to justify the inevitable costs they would suffer in return. For such a strategy to be viable, it would have to be developed from the top down, based on a national space policy that condemns all but retributive attacks on orbital assets and postures the United States as a responsible world leader with the moral authority to protect the common operating environment of orbital space. The final chapter of this monograph offers a way forward for organizing the research needed to inform the development of such a strategy.

CHAPTER FIVE

A Way Forward

This monograph has argued that first-strike stability in space appears to be eroding and that the United States should take concerted action to strengthen that stability by developing a strategy to deter future adversaries from attacking U.S. space systems. Space stability is a fundamental U.S. national security interest. War in space would likely be costly for the United States, even if it were to "win" such a conflict and achieve dominance of that domain. Therefore, U.S. space policies and strategies would better serve the public interest if they were explicitly crafted to deter such conflicts while retaining capabilities to win them in the event of deterrence failure. To support these arguments, the monograph assessed historical shifts in first-strike stability in space and estimated where the thresholds of deterrence failure lie today. It examined the principles of deterrence in the context of the unique operating environment of orbital space to identify what challenges lay before the United States in deterring attacks on its space systems. Finally, it advocated the development of a national space deterrence strategy to meet those challenges.

In doing so, this monograph sketched the broad outlines of a comprehensive space deterrence regime. It must be emphasized, however, that the strategic framework provided here is little more than an empty template for further research. More work would be needed to evaluate which of the various options listed here are both viable and affordable and in what combination they would best support a reliable strategy. Such work would consist of an integrated analysis looking at the space deterrence problem as a complex system and examining the consequences of alternative courses of action, both by the United States and by its most likely potential adversaries, across a range of scenarios. Insights gained from such an examination would inform further analysis to determine near- and far-term approaches for achieving the optimal mix of policies, strategies, and systems for establishing the most effective and affordable deterrence regime. This would entail a broad analysis integrating technical assessments with expert judgment.

Planners would need to gather a good deal of information for this effort, but much of it is available from intelligence sources or has already been developed in previous studies. Among the first steps would be to determine the degree of risk that U.S. space systems currently face and what risks are looming in the foreseeable future. Answering these questions would begin with a thorough assessment of which countries currently have the capabilities to attack which U.S. space systems and what those capabilities are. Beyond that, which capabilities are these countries seeking to develop, and what are their prospects for success in what time frames? Along with the foregoing questions, one would need to survey what capabilities exist, or are currently being developed, to defend U.S. space systems, and which future concepts show the most promise in what time frames. Both passive and active defenses would need to be examined in terms of efficacy and affordability on both government and commercial systems. Answers to these questions would inform an analysis of the potential impacts of attacks on U.S. space capabilities by current and future threat systems. Similar surveys of ongoing efforts and future concepts for developing capabilities for rapid replenishment, dispersal, alternative terrestrial support, and improvements in SSA would also be needed.

On the other side of the cost-benefit calculus, such a study would undertake a fuller investigation into ways in which the United States could punish future enemies for attacks on its space systems. This portion of the study would begin with a close examination of the nature and extent of the international taboo on space warfare. To what extent do people in the United States and other countries believe that outer space is or should be a sanctuary from armed conflict? To what extent can U.S. policies influence such attitudes and bring them to bear on the disadvantage of space aggressors? What are the potential costs and benefits of attempting to shape world opinion in this manner? Whether or not such a taboo exists, how can the United States make its threats of retribution more potent and credible across various levels of confrontation and conflict? What are the escalation risks? Are they acceptable? How should they be managed?

A holistic assessment would require bringing a wide range of analytical methods to bear on this problem. Crisis-gaming and war-gaming would be essential tools for exploring the dynamics of deterrence and stability across a range of scenarios and levels of conflict. Risk analyses and engineering assessments would play important roles in understanding degrees of vulnerability and determining the most promising approaches for mitigating them. Other aspects would include an examination of space law and consultation with other space experts in the U.S. analytical community and elsewhere. Ultimately, having gathered the findings of the surveys, assessments, and analyses, planners would be able to refine and further develop the comprehensive space deterrence strategy outlined in this monograph and offer recommendations for its implementation.

Bibliography

Adams, Karen Ruth, "Attack and Conquer? International Anarchy and the Offense-Defense-Deterrence Balance," *International Security*, Vol. 28, No. 3, Winter 2003–2004.

Air Force Space Command, *Air Force Space Command Strategic Master Plan: FY06 and Beyond*, Peterson AFB, Colo., October 1, 2003.

Behling, Thomas G., "Ensuring a Stable Space Domain for the 21st Century," *Joint Force Quarterly*, No. 47, 4th Quarter 2007, pp. 105–108. As of November 3, 2009:
http://www.ndu.edu/inss/Press/jfq_pages/editions/i47/24.pdf

Betts, Richard K., *Nuclear Blackmail and Nuclear Balance*, Washington, D.C.: Brookings Institution Press, 1987.

Brodie, Bernard, ed., *The Absolute Weapon: Atomic Power and World Order*, New York: Harcourt Brace, 1946.

———, *Strategy in the Missile Age*, Santa Monica, Calif.: RAND Corporation, 1959. As of November 3, 2009:
http://www.rand.org/pubs/commercial_books/CB137-1/

Butterworth, Robert, "Fight for Space Assets, Don't Just Deter," Policy Outlook, Washington, D.C.: George C. Marshall Institute, November 2008. As of November 3, 2009:
http://www.marshall.org/article.php?id=614

Chun, Clayton K. S., *Shooting Down a "Star": Program 437, the US Nuclear ASAT System and Present-Day Copycat Killers*, Center for Aerospace Doctrine Research and Education Paper No. 6, Maxwell AFB, Ala.: Air University Press, April 2000. As of November 3, 2009:
http://handle.dtic.mil/100.2/ADA377346

Commission to Assess United States National Security Space Management and Organization, *Report of the Commission to Assess United States National Security Space Management and Organization*, submitted to the House Armed Services Committee, Washington, D.C., January 11, 2001.

Deblois, Bruce M., "Space Sanctuary: A Viable National Strategy," *Airpower Journal*, Vol. 12, No. 4, Winter 1998, pp. 41–57. As of November 3, 2009: http://www.airpower.maxwell.af.mil/airchronicles/apj/apj98/win98.html

———, ed., *Beyond the Paths of Heaven: The Emergence of Space Power Thought by the School of Advanced Airpower Studies*, Maxwell AFB, Ala.: Air University Press, September 1999.

Defense Advanced Research Projects Agency, Tactical Technology Office, "System F6 Program," Web page, undated. As of December 8, 2009: http://www.darpa.mil/tto/programs/systemf6/

Dolman, Everett C., *Astropolitik: Classical Geopolitics in the Space Age*, London: Frank Cass, 2002.

Freedman, Lawrence, *Deterrence*, Cambridge, UK: Polity Press, 2004.

George, Alexander L., and Richard Smoke, *Deterrence in American Foreign Policy: Theory and Practice*, New York: Columbia University Press, 1974.

Glaser, Charles L., *Analyzing Strategic Nuclear Policy*, Princeton, N.J.: Princeton University Press, 1990.

Gray, Colin S., and John B. Sheldon, "Spacepower and the Revolution in Military Affairs: A Glass Half Full," in Peter L. Hays, James M. Smith, Alan R. Van Tassel, and Guy M. Walsh, eds., *Spacepower for a New Millennium*, New York: McGraw-Hill, 2000, pp. 239–257.

Gruen, Adam L., "Manned Versus Unmanned Space Systems," in R. Cargill Hall and Jacob Neufeld, eds., *The U.S. Air Force in Space: 1945 to the 21st Century*, Proceedings of the Air Force Historical Foundation Symposium, Andrews AFB, Md., September 21–22, 1995, Washington, D.C.: Air Force History and Museums Program, 1998, pp. 67–75.

Hall, R. Cargill, *Military Space and National Policy: Record and Interpretation*, Washington, D.C.: George C. Marshall Institute, 2006.

Hall, R. Cargill, and Jacob Neufeld, eds., *The U.S. Air Force in Space: 1945 to the 21st Century*, Proceedings of the Air Force Historical Foundation Symposium, Andrews AFB, Md., September 21–22, 1995, Washington, D.C.: Air Force History and Museums Program, 1998.

Harrison, Roger G., Darin R. Jackson, and Collin G. Shackelford, *Space Deterrence: The Delicate Balance of Risk*, Colorado Springs, Colo.: Eisenhower Center for Space and Defense Studies, April 2009.

Hays, Peter L., James M. Smith, Alan R. Van Tassel, and Guy M. Walsh, eds., *Spacepower for a New Millennium*, New York: McGraw-Hill, 2000.

Johnson, David E., Karl P. Mueller, and William H. Taft V, *Conventional Coercion Across the Spectrum of Operations: The Utility of U.S. Military Forces in the Emerging Security Environment*, Santa Monica, Calif.: RAND Corporation, MR-1494-A, 2002. As of November 3, 2009:
http://www.rand.org/pubs/monograph_reports/MR1494/

Kent, Glenn A., *Concepts of Operations: A More Coherent Framework for Defense Planning*, Santa Monica, Calif.: RAND Corporation, N-2026-AF, 1983. As of November 3, 2009:
http://www.rand.org/pubs/notes/N2026/

———, *A Framework for Defense Planning*, Santa Monica, Calif.: RAND Corporation, R-3721-AF/OSD, 1989. As of November 3, 2009:
http://www.rand.org/pubs/reports/R3721/

Kent, Glenn A., and William Simons, *A Framework for Enhancing Operational Capabilities*, Santa Monica, Calif.: RAND Corporation, R-4043-AF, 1991. As of November 3, 2009:
http://www.rand.org/pubs/reports/R4043/

Kent, Glenn A., and David E. Thaler, *First-Strike Stability: A Methodology for Evaluating Strategic Forces*, Santa Monica, Calif.: RAND Corporation, R-3765-AF, 1989. As of November 3, 2009:
http://www.rand.org/pubs/reports/R3765/

Kreppon, Michael, testimony before the Strategic Forces Subcommittee, House Armed Forces Committee, March 10, 2009.

Lewis, Leslie, James A. Coggin, and C. Robert Roll, Jr., *The United States Special Operations Command Resource Management Process: An Application of the Strategy-to-Tasks Framework*, Santa Monica, Calif.: RAND Corporation, MR-445-A/SOCOM, 1994. As of November 3, 2009:
http://www.rand.org/pubs/monograph_reports/MR445/

Lupton, David E., *On Space Warfare: A Space Power Doctrine*, Maxwell AFB, Ala.: Air University Press, September 1989.

MacDonald, Bruce W., *China, Space Weapons, and U.S. Security*, Washington, D.C.: Council on Foreign Relations, September 2008.

———, testimony before the Strategic Forces Subcommittee, House Armed Forces Committee, March 18, 2009.

McDougall, Walter A., *The Heavens and the Earth: A Political History of the Space Age*, Baltimore, Md.: Johns Hopkins University Press, 1997.

Mearsheimer, John J., *Conventional Deterrence*, Ithaca, N.Y.: Cornell University Press, 1983.

Morgan, Forrest E., Karl P. Mueller, Evan S. Medeiros, Kevin L. Pollpeter, and Roger Cliff, *Dangerous Thresholds: Managing Escalation in the 21st Century*, Santa Monica, Calif.: RAND Corporation, MG-614-AF, 2008. As of November 3, 2009:
http://www.rand.org/pubs/monographs/MG614/

National Security Decision Memorandum 345, "U.S. Anti-Satellite Capabilities," January 18, 1977. As of January 15, 2009:
http://www.ford.utexas.edu/library/document/nsdmnssm/nsdm345a.htm

Office of Science and Technology Policy, Executive Office of the President, *U.S. National Space Policy*, Washington, D.C.: White House, August 31, 2006.

Office of the Under Secretary of Defense (Comptroller), *Department of Defense Budget Fiscal Year 2009, RTD&E Programs (R-1)*, Washington, D.C., February 2008. As of November 3, 2009:
http://www.defenselink.mil/comptroller/defbudget/fy2009/fy2009_r1.pdf

Peebles, Curtis, *High Frontier: The United States Air Force and the Military Space Program*, Washington, D.C.: Air Force History and Museums Program, 1997.

Perry, William J., James R. Schlesinger, Harry Cartland, John Foster, John Glenn, Mortin Halperin, Lee Hamilton, Fred Ilke, Keith Payne, Bruce Tarter, Ellen Williams, and James Woolsey, *America's Strategic Posture: The Final Report of the Congressional Commission on the Strategic Posture of the United States*, Washington, D.C.: United States Institute of Peace Press, 2009.

Preston, Bob, Dana J. Johnson, Sean J. A. Edwards, Michael Miller, and Calvin Shipbaugh, *Space Weapons, Earth Wars*, Santa Monica, Calif.: RAND Corporation, MR-1209-AF, 2002. As of November 3, 2009:
http://www.rand.org/pubs/monograph_reports/MR1209/

Quester, George H., *Deterrence Before Hiroshima: The Airpower Background of Modern Strategy*, New Brunswick, N.J.: Transaction Books, 1986.

Schelling, Thomas C., *The Strategy of Conflict*, Cambridge, Mass.: Harvard University Press, 1960.

———, *Arms and Influence*, New Haven, Conn.: Yale University Press, 1966.

Schrader, John Y., Leslie Lewis, William Schwabe, C. Robert Roll, Jr., and Ralph Suarez, *USFK Strategy-to-Task Resource Management: A Framework for Resource Decisionmaking*, Santa Monica, Calif.: RAND Corporation, MR-654-USFK, 1996. As of November 3, 2009:
http://www.rand.org/pubs/monograph_reports/MR654/

Schriever, Bernard A., "Military Space Activities: Recollections and Observations," in R. Cargill Hall and Jacob Neufeld, eds., *The U.S. Air Force in Space: 1945 to the 21st Century*, Proceedings of the Air Force Historical Foundation Symposium, Andrews AFB, Md., September 21–22, 1995, Washington, D.C.: Air Force History and Museums Program, 1998, pp. 10–18.

Sheldon, John B., *Space Power and Deterrence: Are We Serious?* Washington, D.C.: George C. Marshall Institute, November 2008.

Smoke, Richard, *National Security and the Nuclear Dilemma: An Introduction to the American Experience*, New York: Random House, 1984.

Stares, Paul B., *The Militarization of Space: U.S. Policy, 1945–1984*, Ithaca, N.Y.: Cornell University Press, 1985.

Treaty on Principles Governing the Activities of States in the Exploration and Use of Outer Space, Including the Moon and Other Celestial Bodies, October 10, 1967. As of February 25, 2010: http://www.oosa.unvienna.org/oosa/SpaceLaw/outerspt.html

U.S. Air Force, *Counterspace Operations*, Air Force Doctrine Document 2-2.1, Washington, D.C., August 2, 2004.

———, *Space Operations*, Air Force Doctrine Document 2-2, Washington, D.C., November 27, 2006.

U.S. Department of Defense, *Annual Report to Congress: Military Power of the People's Republic of China, 2007*, Washington, D.C.: Office of the Secretary of Defense, 2007. As of November 3, 2009: http://www.defenselink.mil/pubs/china.html

U.S. Joint Chiefs of Staff, *Space Operations*, Joint Publication 3-14, Washington, D.C., January 6, 2009.

United Nations General Assembly Resolution 1884 (XVIII), on the question of general and complete disarmament, October 17, 1963. As of November 3, 2009: http://www.un-documents.net/a18r1884.htm

Warner, Edward L. III, and Glenn A. Kent, *A Framework for Planning the Employment of Air Power in Theater War*, Santa Monica, Calif.: RAND Corporation, N-2038-AF, 1984. As of November 3, 2009: http://www.rand.org/pubs/notes/N2038/

Wilson, Tom, "Threats to United States Space Capabilities," prepared for the Commission to Assess United States National Security Space Management and Organization, 2000. As of January 15, 2009: http://www.fas.org/spp/eprint/article05.html

Worden, Simon P., "Space Control for the 21st Century: A Space 'Navy' Protecting the Commercial Basis of America's Wealth," in Peter L. Hays, James M. Smith, Alan R. Van Tassel, and Guy M. Walsh, eds., *Spacepower for a New Millennium*, New York: McGraw-Hill, 2000, pp. 225–238.

Ziegler, David W., "Safe Heavens: Military Strategy and Space Sanctuary," in Bruce M. DeBlois, ed., *Beyond the Paths of Heaven: The Emergence of Space Power Thought by the School of Advanced Airpower Studies*, Maxwell AFB, Ala.: Air University Press, September 1999.